HUMAN-COMPUTER INTERACTION, 1998, Volur
Copyright © 1998, Lawrence Erlbaum Associates,

Introduction to This Special Issue on Experimental Comparisons of Usability Evaluation Methods

Gary M. Olson
University of Michigan

Thomas P. Moran
Xerox Palo Alto Research Center

The field of human–computer interaction has been evolving methods for ensuring the functionality and usability of end-user systems. Practitioners turn to researchers for help in choosing among these methods. Criteria such as efficacy, efficiency, simplicity, timeliness, and cost have been proposed as relevant bases for such choices (J. S. Olson & Moran, 1996). Comparative studies of methods have gained widespread attention and are often used to guide practice as well as the education of HCI professionals.

At the CHI'95 Conference there was a panel titled *Discount or Disservice? Discount Usability Analysis: Evaluation at a Bargain Price or Simply Damaged Merchandise?* Wayne Gray, who organized the panel, presented a very controversial critique of studies that had evaluated various usability evaluation methods (UEMs). The level of interest in this discussion led Gray to propose a review article that dealt with the issues in a more systematic fashion. We encouraged him to do this.

The resulting article, written by Gray and his collaborator Marilyn Salzman, conducted an in-depth review of a series of influential studies that used experimental methods to compare a variety of UEMs. Gray and Salzman's analysis was framed using Cook and Campbell's (1979) well-known discussion of various forms of validity. They used this to evaluate numerous details of these comparative studies, and they concluded that the studies fell far

short on the criteria by which good experimental studies are designed and interpreted.

The article, needless to say, is controversial. The field of HCI is characterized by a diversity of methodological and theoretical approaches, and there is
also the classic tension between the rigor of the pure researcher and the
pragmatics of the practitioner. We knew that readers from this variety of traditions would receive Gray and Salzman's arguments quite differently. We
sought a wide spectrum of reviewers for the article, and from this review process we developed a plan to publish the article in a special issue along with a set
of commentaries from a wide range of contributors to the field of HCI.

There are many important issues raised in this material, and we invite the
reader to plunge in. There is little disagreement among Gray, Salzman, and
the commentators that the studies reviewed have shortcomings. Not surpris-
ingly, there is extensive discussion about the relationship between research on
design methods and the actual practice of design. The ensuing discussion en-
gages such foundational issues as:

- What is meant by usability?
- How are UEMs learned, used, and interpreted in the course of prac-
  ticing design?
- How might research be used to help choose among UEMs?

An important thread in this discussion is the observation, made by a number of the participants, that particular UEMs are almost never used in isolation. This of course complicates even further the question of how to compare
and assess such methods.

There is also extensive discussion of experiments as methods for such research. The level of sophistication about experimental methods of the contributors to this special issue is very high, because many of the commentators,
including many who are now practitioners, were trained as experimentalists.
As a result, the discussion about the strengths and limits of experimental methods is very well informed. It is not surprising that most commentators—and
Gray and Salzman as well—agree that single experiments in isolation are not
the right way to think about the issue. The goal of research should be to use
multiple methods, to replicate and extend key findings, and in general to em-
ploy a strategy of triangulation to achieve research results with the several
forms of Cook–Campbell validity that all agree are desirable. Of importance,
Gray and Salzman point out in their final reply that such triangulation and in-
tegration requires an analytic framework to guide the synthesis of results. Several participants also point out that it is critical to remember that all research
methods have standards of excellence. This is made most vividly by Bonnie
John's comment that "case studies are not just small experiments."

Several participants in this discussion raise issues about publication practices in the field of HCI. The discussion spans such topics as conference papers versus journal articles, practice papers versus "academically respectable" research, and the inherent difficulties in reviewing by and for a heterogeneous community. There are a variety of social and organizational pressures for publishing articles of various kinds and for evaluating the suitability of submissions for various forums. This is an issue worthy of explicit discussion, not just here but in meta-sessions usually held by reviewing bodies.

Work on the development of UEMs and their comparative evaluation has only just begun. UEMs are needed, and what makes them effective and adopted by practitioners is an interesting research area. Research on such real-world problems is difficult, and this is probably an ideal problem to be tackled by academic–industrial partnerships, as Gray and Salzman suggest.

REFERENCES

Cook, T. D., & Campbell, D. T. (1979). *Quasi-experimentation: Design and analysis issues for field settings.* Chicago: Rand McNally.

Olson, J. S., & Moran, T. P. (1996). Mapping the method muddle: Guidance in using methods for user interface design. In M. Rudisill, C. Lewis, P. G. Polson, & T. D. McKay (Eds.), *Human–computer interface designs: Success stories, emerging methods, and real world context* (pp. 269–300). San Francisco: Morgan Kaufmann.

ARTICLES IN THIS SPECIAL ISSUE

Gray, W. D., & Salzman, M. C. (1998). Damaged merchandise? A review of experiments that compare usability methods. *Human–Computer Interaction, 13,* 203–261.

Karat, J., Jeffries, R., Miller, J., Lund, A. M., McClelland, I., John, B. E., Monk, A. F., Oviatt, S. L., Carroll, J. M., Mackay, W. E., & Newman, W. N. (1998). Commentaries on "Damaged merchandise?" *Human–Computer Interaction, 13,* 263–323.

Gray, W. D., & Salzman, M. C. (1998). Repairing damaged merchandise: A rejoinder. *Human–Computer Interaction, 13,* 325–335.

HUMAN-COMPUTER INTERACTION, 1998, Volume 13, pp. 203–261
Copyright © 1998, Lawrence Erlbaum Associates, Inc.

Damaged Merchandise?
A Review of Experiments
That Compare Usability
Evaluation Methods

Wayne D. Gray and Marilyn C. Salzman
George Mason University

ABSTRACT

An interest in the design of interfaces has been a core topic for researchers and practitioners in the field of human–computer interaction (HCI); an interest in the design of experiments has not. To the extent that reliable and valid guidance for the former depends on the results of the latter, it is necessary that researchers and practitioners understand how small features of an experimental design can cast large shadows over the results and conclusions that can be drawn. In this review we examine the design of 5 experiments that compared usability evaluation methods (UEMs). Each has had an important influence on HCI thought and practice. Unfortunately, our examination shows that small problems in the way these experiments were designed and conducted call into serious question what we thought we knew regarding the efficacy of various UEMs. If the influence of these experiments were trivial, then such small

Wayne Gray is a cognitive scientist with an interest in how artifact design affects the cognition required to perform tasks; he has worked in government and industry and currently heads the Human Factors and Applied Cognitive Program at George Mason University. **Marilyn Salzman** is a design and usability engineer for U S WEST Advanced Technologies and a doctoral student in the Human Factors and Applied Cognitive Program at George Mason University. Her interests include human–computer interaction design and evaluation.

CONTENTS

---

problems could be safely ignored. Unfortunately, the outcomes of these experiments have been used to justify advice to practitioners regarding their choice of UEMs. Making such choices based on misleading or erroneous claims can be detrimental—compromising the quality and integrity of the evaluation, incurring unnecessary costs, or undermining the practitioner's credibility within the design team. The experimental method is a potent vehicle that can help inform the choice of a UEM as well as help to address other HCI issues. However, to obtain the desired outcomes, close attention must be paid to experimental design.

1. OVERVIEW

Usability is a core construct in human–computer interaction (HCI). Methods to evaluate the usability of various software packages have been of intense interest to HCI researchers and practitioners alike. Various usability evaluation methods (UEMs) have been created and promoted. The appeal of some UEMs rests on common sense and the persuasiveness of proponents of that UEM. Others are based on case studies or lessons learned and collected from various organizations. Finally, other UEMs have been promoted based on the results of experimental studies designed to compare the effectiveness of two or more UEMs.

This review is limited to the latter sorts of arguments: experimental studies intended to yield objective and generalizable data regarding the utility of one or more UEM. As we show later, the most influential of such experiments suffer from two basic problems. First, there are measurement issues. It is not clear that what is being compared across UEMs is their ability to assess usability. Although something is being measured, it is far from obvious that these measures really reflect sensitivity to usability. Second, there are design issues. The design of many of the experiments is such that neither the data they produce nor the conclusions drawn from the data are reliable or valid.

The implications we draw are of more than *academic* interest. They concern the entire HCI community. The parts of the review that examine the validity of assertions that one UEM is better than another may be primarily of interest to practitioners. The parts that discuss methodological and logical failings in experimental design may be of primary interest to researchers. However, the whole of the review is greater than the sum of its parts. Hence, this article is not simply a review of experimental findings or simply a methodology-oriented discussion of experimental failings.

Both authors are currently academics but formerly practitioners. We brought to our review the practitioner's disdain of hair-splitting academic arguments. However, as we warmed to our topic we found, time and again, numerous small problems in how the research was designed and conducted. If these small problems had only small effects on the interpretation of that research, we would have ignored them. Unfortunately, these small problems had large effects on what we could legitimately learn from the experiments. Cumulatively, they called into serious question what we thought we knew regarding the efficacy of various UEMs. We therefore ask our reader's indulgence in following us in the explanation of these small problems. This is not an exercise in hair-splitting. It is an attempt to convince you that much of what you thought you knew about UEMs is potentially misleading.

2. INTRODUCTION

UEMs are used to evaluate the interaction of the human with the computer for the purpose of identifying aspects of this interaction that can be improved to increase usability. They typically come into play sometime after needs assessment and before beta testing (see Olson & Moran, 1996, for a general discussion of where UEMs fit in the software development life cycle). UEMs can be categorized as analytic or empirical. *Analytic UEMs* include techniques such as Heuristic Evaluation, (Nielsen & Molich, 1990), Cognitive Walkthrough (C. Lewis & Polson, 1992; Wharton, Rieman, Lewis, & Polson, 1994), guidelines (e.g., Smith & Mosier, 1986), GOMS (Card, Moran, & Newell, 1983; John &

Kieras, 1996a, 1996b), and others. *Empirical UEMs* include a wide range of methods and procedures that are often referred to simply as user testing.

Our focus in this review is not on UEMs per se but on the studies that were intended as experimental manipulations to compare and contrast UEMs.[1] The purpose of these studies was to provide guidance to practitioners regarding the effectiveness of various UEMs. Our review tightly focuses on the design of the studies: what UEMs were used, how usability was measured, who the participants were, what the participants did, and so on. Our intent is threefold. First, we draw conclusions regarding whether the design of the study supports the claims that were made. Second, when the claims do not follow from the experimental manipulation, we identify the source of the problem. Third, over the body of studies reviewed, we uncover common problems and make suggestions for how experimental research in HCI can be improved.

In the cause of usability, doing something is almost always better than doing nothing. However, for HCI practitioners, making choices based on misleading or erroneous claims can be detrimental—compromising the quality and integrity of the evaluation, incurring unnecessary costs, or undermining the practitioner's credibility within the design team. For UEMs to provide useful guidance to the design process, comparison studies must be conducted that delineate the trade-offs—the advantages and disadvantages—of each method. Such experiments cannot be conducted quickly or easily. They require a substantial commitment of time and resources. Necessarily, all such experiments are limited in scope, and these limits must be explicitly acknowledged. Such limits do not mean that guidance to practitioners cannot be forthcoming or that the experimental method does not have a role to play in formulating that guidance. If the power of the experimental approach is to be applied to illuminate the advantages and pitfalls of various UEMs, then broad-brush experiments must be eschewed and a program of multiple, narrowly focused experiments must be embraced.

The plan for this article is as follows. To delimit and focus the scope of our complaints, we begin by discussing the unique role of experiments among empirical methods. We then quickly review four well-known threats to validity in experimental studies (Cook & Campbell, 1979) and discuss a practice that seems widespread in the HCI literature: going beyond the data to provide advice to practitioners. In the main section, we review five papers that have compared UEMs: Jeffries, Miller, Wharton, and Uyeda (1991); Karat, Campbell, and Fiegel (1992); Nielsen, (1992); Desurvire, Kondziela, and Atwood (1992);

---

1. N.B., although our focus is on studies that compare UEMs, we have no reason to believe that the problems we find in these studies are not endemic to other areas of HCI research.

and Nielsen and Phillips (1993). All of these studies have been well cited, all
were published in refereed conference proceedings (four in the prestigious
SIGCHI proceedings and one in the well-regarded People and Computers
Conference proceedings), and all have had enough time for fuller reports to
appear in the literature (none have). Having provided a detailed introduction
of threats to validity and having then used this framework to organize a detailed review of five studies, in the Observations and Recommendations section we suggest ways that future HCI experiments can be designed to avoid
these threats.

3. THE UNIQUE ROLE AND BURDEN OF EXPERIMENTS IN EMPIRICAL STUDIES OF HUMAN–COMPUTER INTERACTION

Despite the plethora of alternatives, the traditional experimental approach
continues to allure researchers. It is important to understand why. Simply put,
"the unique purpose of experiments is to provide stronger tests of *causal* hypotheses than is permitted by other forms of research" (Cook & Campbell,
1979, p. 83).

A well-conducted, valid experiment permits us to make strong inferences
regarding two important issues: (a) cause and effect and (b) generality. Experiments are conducted to determine the effect of some independent variable
(e.g., type of UEM) on some dependent variable (DV) (e.g., the number of usability problems uncovered). Experiments permit us to go beyond noting cor-
relation to inferring causality. For example, if using UEM-A, Group 1
identifies more usability problems than Group 2 (that used UEM-B), we can
infer that it was the use of UEM-A, and not some other factor, that *caused*
Group 1 to find more problems (*effect*) than Group 2.

Generality is as important to experimenters as causality. Is the effect found
limited to the exact circumstances of the study or can it be generalized to other
circumstances? If the study were conducted at NYNEX, could it be generalized to IBM? If the study used usability specialists with 3 years of experience,
could it be generalized to specialists with 10 years of experience? Can the solutions to usability problems discovered by experimental methods be generalized to problems observed in usability testing?

The inference of cause and effect and the claim of generality is the essence
of the experimental method and the reason for its continued attraction. However, few studies are generalizable across all times and places. Many have one
or more limits in their claim to causality. When researchers become aware of
problems and limits imposed by the conditions of their studies, it is their responsibility to call these limits to the attention of the reader and to explicitly
circumscribe the claims made.

4. THREATS TO THE VALIDITY OF EXPERIMENTAL STUDIES

What sorts of things should we look at to decide, first, whether a study was well done and, second, how far we can generalize the findings? These issues are often referred to as different types of validity. Although there are different breakdowns of validity, we consider the four discussed by Cook and Campbell (1979): statistical conclusion validity, internal validity, construct validity, and external validity. Additionally, we discuss the practice of using the discussion or conclusion section to go beyond what was investigated in the experiment to provide advice to practitioners. Each threat to validity is multifaceted, and we focus on those facets that are most relevant to HCI research. These facets become major themes in our review of UEM studies.

4.1. Cause–Effect Issues

Statistical conclusion validity and internal validity are concerned with drawing false positive or false negative conclusions. Succinctly put, statistical conclusion validity helps us establish whether the independent variable is related to the DV. If there is a relation, then the question for *internal validity* is whether we can conclude that the independent variable (the treatment) caused the observed change in the DV (what we measured) or whether both variables are simply correlated and the observed changes were caused by a third, unnoted variable.

Statistical Conclusion Validity

Statistical conclusion validity is a realm well covered by statistics textbooks. The issues of particular concern for UEM studies include low statistical power, random heterogeneity of participants, and doing too many comparisons.

Low statistical power and random heterogeneity of participants might be regarded as two sides of the same coin. Low statistical power may cause true differences not to be noticed; random heterogeneity of participants may cause noticed differences not to be true. Potential solutions to these problems are to increase the number of participants per group and to consider group differences in the context of individual differences (variability).

Many UEM researchers use simple descriptive statistics (such as averages, percentages, and tallies) and tend to rely on *eyeball tests*[2] to determine if appar-

---

2. *Eyeball test* refers to the practice of looking at the data (eyeballing it) and deciding by intuition that differences between two raw numbers, percentages, or means are real.

ent differences are real. Unfortunately, avoiding statistics does not avoid problems with statistical conclusion validity. When the power of the findings is too low to use a statistical test, the sample size may be too low to provide a stable estimate of an effect. In many cases of low sample size, effects due to the random heterogeneity of participants may be greater than the systematic effects of the treatment (e.g., type of UEM or type of expertise). Another way of phrasing this problem is to think of one or more of the participants as *Wildcards*—that is, people who are significantly better or worse than average and whose performance in the conditions of the study do not reflect the UEM but reflect their Wildcard status. The simplest solution to this *Wildcard effect*[3] is to randomly assign more usability interface specialists (UISs)[4] to each UEM treatment. The Wildcard effect is less likely to influence results when the Wildcard is 1 or 2 UISs in a group of 10 or 20 instead of 1 in a group of 3. The checks on the Wildcard effect are statistics such as the *t* test or analysis of variance that compare the hypothesized systematic effect due to treatment (i.e., UEM) to the presumably random effect due to participant. If the treatment effect is enough bigger than the Wildcard effect, then the difference between the two groups can be attributed to the UEMs and not to their Wildcards.

In addition to low power and random heterogeneity of variance (the Wildcard effect), the way in which comparisons are selected can pose a threat to validity. Experimental design handbooks, such as Keppel and Saufley (1980), warn that "unplanned comparisons are considered 'opportunistic' in the sense that they can capitalize on chance factors" (pp. 140–141). They urge researchers to "take precautions against becoming overly 'zealous' in declaring that a particularly attractive difference" (p. 141) is real. The lure of interpreting these "particularly attractive differences" seems to be great in UEM experiments. In judging the potential problem posed by multiple comparisons, the important distinction is between comparisons that are chosen after examining the data versus those that follow from the logic of the experimental design.

Problems due to low power, random heterogeneity of participants, and unplanned comparisons can be controlled by the use of standard statistical techniques. Such techniques attempt to ensure that the random effect due to different participants is significantly less than the treatment effect due to the

---

3. The term *Wildcard* is meant to apply to variability among novices as well as among experts. (Note that some decks, or studies, may have more than one Wildcard.) Our attempt at creating a mnemonic label for this effect should not obscure the basic statistical concern with the testing of group differences against participant (subject) variability.

4. N.B., the studies we reviewed often define two types of evaluators: one whose primary responsibility is human factors or interface issues and another whose primary responsibility is designing or writing code. For consistency's sake we refer to these two as UISs or software engineers (SWEs).

experimental manipulation. They also provide ways of doing multiple planned and unplanned comparisons while mitigating the capitalization on chance factors.

Experiments Versus User Testing. Statistical conclusion validity is more of an issue for those who would conduct experiments than for those involved in user testing. Because user testing assumes so many of the trappings of the experimental method it may seem reasonable to accept standards that are appropriate for user testing as appropriate for experimental research. However, this should not be done. A major distinction between the two is illustrated by the number of participants required to obtain meaningful data. *Experimental* results have shown that for *user testing,* only a few participants are needed to identify problems and even fewer participants are needed to identify severe problems (Virzi, 1992; although see J. R. Lewis, 1994, for a counter argument). Similar studies were performed by Nielsen (1992; Nielsen & Molich, 1990) and resulted in the recommendation to use three to five UISs for Heuristic Evaluation.

The distinction between standards appropriate for the usability lab and those appropriate for research is well illustrated by the difference between the methods and the recommendations of this series of studies. Support for the use of a handful of users or UISs in usability testing comes from experiments that used large numbers of participants (e.g., Virzi performed three studies with 12, 20, and 20 participants; Nielsen and Molich used 34 participants; and Nielsen used three groups of 31, 19, and 14 participants).

Internal Validity

Statistical conclusion validity establishes that there are real differences between groups; internal validity concerns whether these differences are causal as opposed to correlational (there might be a third, unknown variable that is responsible for the changes). Unfortunately, there is no simple test for internal validity. There are, however, well-established issues researchers need to consider. Here we consider three: instrumentation, selection, and setting (see Cook & Campbell, 1979, for a fuller exposition).

Instrumentation. For UEM studies, instrumentation primarily concerns biases (covert or overt) in how human observers identify or rate the severity of usability problems. Comparing methods or groups is only valid if there is a way of rating the results that does not inappropriately favor one condition over the others.

In several of the UEM studies we reviewed, evaluators were assigned to different UEMs and asked to identify usability problems using that UEM. This approach was threatened by instrumentation problems when either the evaluators or the experimenters were required to identify, classify, or rate the usability problems. For example, if, during the course of the study, the evaluators changed how they identified, classified, or rated problems, then an instrumentation problem existed. Factors that could have resulted in such changes are increased sensitivity to problems, increased experience with the system, and so forth. Another common flaw in the instrumentation of UEM studies is when problem categories defined by one UEM, for example, Heuristic Evaluation, were used by the experimenters to categorize problems found by another UEM, for example, user testing. In this case, the perspective of one UEM was likely to have biased the count or classification of problems found by the other UEM.

The issue is that *you find what you look for* and, conversely, *you seldom find what you are not looking for*. Although steps can be taken to prevent instrumentation problems, the scant discussion of this issue in the UEM literature raises rather than allays our concerns.

Selection.   Selection is a threat "when an effect may be due to the difference between the kinds of people in one experimental group as opposed to another" (Cook & Campbell, 1979, p. 53). In our review we distinguish between general versus specific selection threats. A *general selection* threat refers to a characteristic of the participants that is not directly related to the manipulation of interest. A *specific selection* threat exists when the participants assigned to different groups are unequal in some characteristic (e.g., knowledge or experience) that is directly related to some aspect of the experimental procedures (and is not the intended manipulation).

Setting.   The setting for an experiment may influence its outcomes. Indeed, an interesting research study might involve training consumers in, for example, Cognitive Walkthrough and determining whether their evaluations of home software change as a function of applying the technique in a show room (before purchase), at home (after purchase), or in the usability lab. In this example, *setting* becomes the independent variable (i.e., something that is being manipulated to determine whether it exerts any significant effect on the outcome). In some of the studies we reviewed, the setting covaried with UEM (treatment), type of participant (e.g., UIS vs. SWE), or both. In these cases, differences in setting is a threat to the study's internal validity because it is impossible to determine whether the effect observed was obtained from the treatment, the setting, or the treatment–setting combination.

4.2. Generality Issues

If a study is internally and statistically valid, then we want to know whether the causal relationships found can be generalized to alternative measures of cause and effect as well as across different types of persons, settings, and times. Cook and Campbell (1979) referred to these issues as construct validity and external validity.

Construct Validity

Construct validity divides neatly into two issues: Are the experimenters manipulating what they claim to be manipulating (the *causal construct*) and are they measuring what they claim to be measuring (the *effect construct*)? UEM studies have problems on both of these dimensions.

Causal Construct Validity

In the studies we reviewed, we found several types of threats to construct validity. The first threat is the most basic, as well as the most pervasive: operationalizing, or defining, the UEM. Did the way in which the experimenters conducted UEM-A correspond to the reader's understanding of UEM-A? Additional threats could arise from using only one way of applying a flexible UEM (mono-operation bias), applying the UEM to just one type of software (mono-method bias), or from the interaction of different treatments. In the following we discuss each of these issues in more detail.

Defining the UEM. The development and definition of UEMs has been a dynamic enterprise. In fact, all currently used analytic UEMs have evolved rapidly over recent years. Therefore, it is understandable that the exact definition of any given UEM may have changed over time. Unfortunately, changing definitions while keeping the names the same makes it hard if not impossible to know what UEM is being manipulated and, hence, to compare outcomes from different UEM studies.

Analytic UEMs such as walkthroughs, guidelines, and heuristic evaluations are the UEMs that suffer most from causal construct problems. To facilitate communication and comparison, we use the terminology in Figure 1, which expands on Olson and Moran's (1996) terminology.

The two dimensions captured in Figure 1 are guidelines and scenarios. A scenario exists when evaluators are told to perform a given set of tasks or are asked to evaluate the steps of a task as they would be performed by the user (sometimes a flowchart is given, other times a listing is provided). Guidelines are broadly defined as any list of problems, features, or attributes provided to

Figure 1. Usability Evaluation Methods Terminology Guide.

	Scenario	
Guidelines	No	Yes
None	Expert review	Expert walkthrough
Short list	Heuristic evaluation	Heuristic walkthrough
Long list	Guidelines	Guidelines walkthrough
Information processing perspective	N/A	Cognitive walkthrough

evaluators for the purpose of determining whether any item from this list has been instantiated in the interface.

If experts are simply given an interface and told to evaluate it without being provided specific guidelines or specific scenarios, we call it an expert review (or simply a review if the evaluators are not experts). We reserve Heuristic Evaluation for Nielsen's "discount" technique (Nielsen, 1992, 1993, 1994a, 1994b; Nielsen & Molich, 1990) that, among other things, provides evaluators with a short list of guidelines but no scenario. Having access to a long list of guidelines[5] but no scenario is referred to simply as guidelines.

When experts are simply given a scenario and told to use it in performing their evaluation, this is an expert walkthrough. A heuristic walkthrough provides evaluators with a short list of guidelines with which to identify problems found during the walkthrough. Similarly, a guidelines walkthrough provides evaluators with a long list of guidelines to use during the walkthrough. Finally, Cognitive Walkthrough is reserved for conducting a walkthrough using the techniques derived from CE+ theory (C. Lewis & Polson, 1992; Polson, Lewis, Rieman, & Wharton, 1992; Wharton, Bradford, Jeffries, & Franzke, 1992; Wharton et al., 1994).

From the perspective of causal construct validity, we cannot infer that UEM-A is better than UEM-B unless we are certain that the methods used by the experimenter were, in fact, representative of UEM-A and UEM-B. In the reviews that follow, when the terminology used by the researchers differs from the terminology we use in Figure 1, we point out the difference but continue to use the terminology of the paper being reviewed.

---

5. The short lists used for Heuristic Evaluation typically have less than a dozen guidelines. In contrast, the longer lists have several dozen or more guidelines. For example, one common source of guidelines (Smith & Mosier, 1986) has several hundred.

Mono-Operation and Mono-Method Bias. Adopting Figure 1's terminology does not ensure that different experiments have implemented the same UEM in the same way. Important variations may still exist. For example, in doing a Heuristic Evaluation, members of one group may conduct their evaluations independently and then combine results, whereas members of another may work together as a team. It is important to know (as researchers and practitioners) how changes in the way a UEM is carried out (i.e., *operations*) affect its ability to identify problems.

Mono-*method* bias is a complement to mono-operation bias. Just as there are many differences in how any one UEM can be used, there are many differences in the type of software that is to be evaluated. It is not obvious, for example, that a UEM that is good at finding problems with office automation software can be trusted to identify problems with real-time, safety-critical systems. Practitioners need to know what UEM works best for the type of software they are developing. Researchers need to understand whether and how usability problems vary with the type of software. Mono-method bias may lead us to draw false generalizations concerning both of these issues.

Interaction of Different Treatments: Confounding. In a few of the studies reviewed, two or more UEMs were used by the same set of participants. Such designs raise a threat to causal construct validity due to the possible *interaction of different treatments*. A threat exists because the experience gained by using UEM-A may affect the behavior (judgments or whatever) of participants while using UEM-B. For example, when using UEM-A participants are identifying problems as well as gaining familiarity with the software. Both these UEM-A outcomes (uncovering problems plus gaining familiarity) may be expected to feed into what the participants do and how they view the software later while using UEM-B. What is really being manipulated is not simply UEM-A versus UEM-B but UEM-A versus UEM-A/B; that is, the second treatment is not the pure form of UEM-B but is UEM-B *confounded* by UEM-A.

Effect Construct Validity: Intrinsic Versus Payoff Measures of Usability

An important distinction for UEMs is the difference between *intrinsic* and *payoff* measures of usability (Scriven, 1977):

> If you want to evaluate a tool ... say an axe, you might study the design of the bit, the weight distribution, the steel alloy used, the grade of hickory in the handle, etc., or you might just study the kind and speed of the cuts it makes in the hands of a good axeman. (p. 346)

Empirical UEMs (i.e., most varieties of user testing) may measure performance directly. They are equivalent to Scriven's (1977) study of the "kind and speed of the cuts." Analytic UEMs examine the interface or aspects of the interaction and infer usability problems. This is equivalent to Scriven's study of the bit, weight distribution, steel alloy, and grade of hickory. UEMs such as those of the GOMS family have demonstrated utility for relating intrinsic attributes of the interface to payoffs (or costs) of using the interface (performance outcomes). For example, NGOMSL has been used to predict speed of learning (Bovair, Kieras, & Polson, 1990; Kieras, 1997) and CPM-GOMS (Gray, John, & Atwood, 1993, pp. 282–286) has pinpointed performance problems to specific features of the keyboard layout, screen layout, keying procedures, and system response times. For other UEMs, such as guidelines and Heuristic Evaluation, making this forward inference (from feature to payoff) is not as tight or as obvious.

We find much confusion in the literature concerning the nature and role of the two types of UEMs. Analytic UEMs examine intrinsic features and attempt to make predictions concerning payoff performance. Empirical UEMs typically attempt to measure payoff performance directly (e.g., speed, number of errors, learning time, etc.). When an empirical UEM is used to compare the usability of two different interfaces on some measure(s) of usability (e.g., time to complete a series of tasks) the results are clear and unambiguous: The faster system is the more usable (by that criterion for usability).

An opportunity for problems arises when the outcomes of empirical and analytic UEMs are viewed as equivalent—that is, when empirical UEMs are used in an attempt to identify intrinsic features that caused the observed payoff problem. Empirical UEMs can identify problems, but care must be taken to isolate (e.g., Landauer, 1988) and identify the feature that caused the problem. None of the studies we reviewed report systematic ways of relating payoff problems to intrinsic features; all apparently rely on some form of expert judgment.

Problems of interpretation arise when the number of problems identified by one UEM is compared to those identified by another. When different techniques identify different problems, do the differences represent misses for one UEM or false alarms for the other? When comparing results of an analytic UEM with an empirical UEM, is a feature identified by intrinsic evaluation a usability problem even if it has no effect on performance?

We believe that *effect construct validity* is the single most important issue facing HCI researchers and practitioners. We do not dwell on this issue during our review of the five studies; however, we return to this topic afterward (see Section 6.1).

External Validity

External validity concerns generalizing *to* particular target persons, settings, and times and generalizing *across* types of persons, settings, and times (Cook & Campbell, 1979). The distinction is between generalizing to a population versus across subpopulations.

For example, Karat et al. (1992) brought "together a group that had the composition of a development team (developers and architects, UI specialists, and appropriate support personnel) working with end users" (C. -M. Karat, personal communication, June 1, 1995). As intended by the experimenters, these results *generalize to* the heterogeneous population used by IBM development teams. However, readers who attempt to *generalize across* Karat et al.'s sample to groups composed of just UISs or just end users or just developers would be making an error of external validity. Again, the distinction is one of generalizing to a similarly heterogeneous population (okay) versus generalizing across the subpopulations (not okay).

Claims that exceed the scope of the *settings* and *persons* that the experiment can generalize *to* or *across* are said to lack external validity. The difference between the exact settings and persons used in the experiment and the wider range of settings and persons to which the experimenter seeks to generalize is a constant source of tension in reporting experimental results. When reporting results, researchers should attempt to balance grand claims against explicitly stated limitations. An example of such a qualification might be: "Although we believe that UEM-A can be used by any UIS with no special training, we must note that the 12 UISs used in this study all had PhDs in Cognitive Psychology and had been involved, for the past 5 years, in a full-time effort to develop UEM-A." Unfortunately, although broad claims are rampant, explicitly stated limitations and caveats are rare.

4.3. Conclusion Validity

Are the claims made by the authors consistent with the results or do the claims follow from what was done? Our concern in this section is not with claims that are invalid due to one of the four Cook and Campbell (1979) validity problems. In the case of a problem with one of the first three types of validity (statistical conclusion validity, internal validity, or construct validity), presumably the research was designed to address that claim but ran into trouble for other reasons (that, unfortunately, were not noticed by the researchers). Likewise, for the fourth type of validity, external validity, presumably the claim made is an overgeneralization of a particular finding.

Our concern is with claims made in the discussion or conclusion of the study that were either not investigated in the study or contradicted by the results of the study. The former is *beyond the scope* of the study; whereas the latter is a *contradicted conclusion*. Although we can see no justification for contradicted conclusions, there are some interesting issues involved in claims that go beyond a study's scope.

For example, a reviewer (anonymous, personal communication, December 8, 1996) of an earlier version of this article lambasted us for our

> attempt to interpret everything said in the Conclusions or Discussion section of these papers as conclusions that logically follow from the analyses in the paper. Often statements made in such sections are comments or advice to others who might want to use these results.

There is a tradition in the human factors literature of providing advice to practitioners on issues related to, but not investigated in, an experiment. This tradition includes the clear and explicit separation of *experiment*-based claims from *experience*-based *advice*. Our complaint is not against experimenters who attempt to offer *good advice*. Rather, we are concerned with advice that is offered without the appropriate qualifications. Experience-based advice needs to be clearly and explicitly distinguished from experiment-based inference. Unless such care is taken, the advice may be understood as *research findings* rather than as the *researcher's opinion*.

4.4. Summary: Threats to Validity of Experimental Studies

Cook and Campbell (1979) discussed four threats to the validity of experimental studies. We have attempted to define these threats within the context of HCI research. In the next section, we examine the validity of five major studies of UEMs and discuss how validity problems threaten the conclusions one can draw based on these studies. We also note occasional contradicted conclusions, as well as experience-based advice that is not explicitly distinguished from experiment-based inference.

5. THREATS TO THE VALIDITY OF FIVE INFLUENTIAL UEM STUDIES

The body of this review examines five influential UEM studies. In order of publication, these are Jeffries et al. (1991), Karat et al. (1992), Nielsen (1992), Desurvire et al. (1992), and Nielsen and Phillips (1993). Although it is arguable whether these are the *most* influential UEM studies, it seems beyond dispute that they have been *very* influential. Our arguments for influence are based on informal surveys of the citation of these works in the proceedings of recent HCI conferences, journals, textbooks, and books. We recognize that many other noteworthy UEM studies have been reported in the literature. In fact, in an earlier version of this article we included 11 extended reviews. However, for brevity, we have limited our discussion to a core set of reviews.[6]

Each review has three parts:

> *Overview:* A description of the goals of the experiment and a brief summary of the study's design and methodology.
>
> *Validity Issues:* An examination of the validity problems that limit our acceptance of the claims made by the researchers.
>
> *Summary:* Our summary of the major problems with validity as well as what we believe can be safely concluded from the study as conducted.

As a type of advanced organizer, we refer the reader to Figure 2 and Figure 3. For each study, Figure 2 indicates potential problems with how comparisons were made, the number of participants per group, and the type of statistics used (or the lack thereof). (Note that the individual entries in Figure 2 are explained in detail as the study is discussed). Figure 3 summarizes our judgment regarding each study's most severe validity problems. In addition, we refer you to Appendices A-1 through A-5 for detailed lists of the claims made by the researchers for each study as well as the problems that threaten the validity of those claims.

5.1. Jeffries, Miller, Wharton, and Uyeda (1991)

Overview

Jeffries et al. (1991) compared four UEMs that they called Heuristic Evaluation, Cognitive Walkthrough, guidelines, and user testing. By our classification, the first is an expert review rather than a Heuristic Evaluation; however, the others appear to be good exemplars of their categories (see Figure 1). All groups assessed the usability of HP-VUE™ "a visual interface to the UNIX operating system" (p. 120). There were 4 UISs in the Heuristic Evaluation group,

---

6. In the more extended version our article, we characterized four of our current five studies as the most influential UEM-comparison studies. An additional four studies were listed as less influential UEM comparisons. The final three were reviewed as UEM-expertise studies. Whereas none of the readers of these versions felt inclined to add studies to our list, many urged us to shorten the article by reducing the studies reviewed to a core set. Hence, we kept our original four UEM-comparison studies and added what we believe is the most influential of the UEM-expertise studies (i.e., Nielsen, 1992).

Figure 2. Summary of information discussed in Statistical Conclusion Validity.

Experiment	Number of Comparisons	Participants Per Group	Type of Statistics	
			Participant Variability Considered?	Eyeball or Other
Jeffries et al. (1991)	Potential problem	4-3-3-6(UT)	No	Most
Karat et al. (1992)	Okay	6-6-6-6-6-6	No	Most
Nielsen (1992)[a]	Potential problem	31-19-14	Some	Many
Desurvire et al. (1992)	Potential problem	3-3-3-3-3 each; 18(UT)	No	All
Nielsen and Phillips (1993)	Potential problem	12-10-15-19-20(UT)	Variance reported	Most

Note. UT = user testing.
[a]Data summary based on the experimental part of the paper only.

Figure 3. Problems by validity types across studies.

Experiment	Erroneous Claims Due to Problems With:			Claims That Exceed Scope of Study Due to:	
	Statistical Conclusion Validity	Internal Validity	Construct Validity	External Validity	Conclusion Validity
Jeffries et al. (1991)	X	X	X	X	X
Karat et al. (1992)	X		X	X	X
Nielsen (1992)	X	X	X		X
Desurvire et al. (1992)	X	X	X	X	X
Nielsen and Phillips (1993)	X	X	X	X	X

one team of 3 SWEs in the guidelines group, one team of 3 SWEs in the Cognitive Walkthrough group, and 6 "regular PC users" who were not familiar with UNIX in the user testing group. The user testing was conducted by a human factors expert (UIS) practiced in user testing.

Jeffries et al.'s research goals were to determine (a) the interface problems the UEMs best detected, (b) the relative costs and benefits of each UEM, and (c) who (developers or UI specialists) could use the UEMs more effectively, all in a real-world setting. Based on the outcomes of the study, they drew inferences about each of these issues. However, a number of threats to validity limit the strength of their claims.

Statistical Conclusion Validity

Sample Size.   With 3 to 6 participants per group, this study suffered from low statistical power and its concomitant problem of random heterogeneity of participants—that is, just the sort of situation in which the Wildcard effect is likely to have occurred. With so few people per group, small variations in individual performance could have had a large influence on the stability of measures of group differences, making us unable to draw valid inferences about one UEM versus another.

There is a second and more subtle problem here as well. Participants in some groups worked as teams, making teams the unit rather than individual participants. Hence, Figure 2, which shows 4-3-3-6(UT) user testing participants per group, actually provides a lenient count for sample size. A more accurate count would reflect the number of teams in a group, 4-1-1-1. The Heuristic Evaluation group had four teams (with 1 participant each) because each participant conducted his or her Heuristic Evaluation independently. Participants in the guidelines and Cognitive Walkthrough groups worked as teams, providing a sample size of one for each those groups. It also appears that we should think of the user testing condition as composed of one team because a single UIS conducted the user tests.

Statistics Used.   No statistical analyses were performed, and the variability of individual performance was not accounted for. For example, the claim that "the [H]euristic [E]valuation technique as applied here produced the best results. It found the most problems, including more of the most serious ones, than did any other technique, and at the lowest cost" (p. 123) was based on a comparison of the total number of problems found by each group. Like this claim, all of Jeffries et al.'s claims were based on informal comparisons of totals, percentages, and means for very small sample sizes. In addition, the large number of comparisons (see Figure 2) seems likely to have capitalized on chance factors.

Internal Validity

Selection.   Groups differed in the type of participant assigned to them, creating internal validity problems with selection. A *general selection* problem (see Section 4.1, Internal Validity) existed because UISs composed the Heuristic Evaluation group, whereas the Cognitive Walkthrough and guidelines groups consisted of SWEs. In addition, a *specific selection* problem existed in that 2 of the 3 SWEs in the guidelines group had extensive experience (40 and 20 hr) with the application (HP-VUE™) being evaluated, but participants in the other groups did not. We have no way of knowing whether the finding that "the guidelines evaluation was the best of the four techniques at finding recurring and general problems" (p. 123) was due to the guidelines UEM or due to the experience with HP-VUE that members of the guidelines group had that was, apparently, not possessed by any member of any other group. These general and specific selection threats make it impossible to separate the influence of participant background from the utility of the technique.

Setting.   Random irrelevancies in experimental setting presented an additional threat to internal validity. The Heuristic Evaluation group assessed HP-VUE at their own pace over a 2-week period and, presumably, at their own machines. The user testing group was given 3 hr of HP-VUE training followed by 2 hr of testing. Although the text is not clear, it appears that the guideline and Cognitive Walkthrough groups also completed evaluations in one sitting. Such extreme variations in setting might have affected group performance.

Instrumentation.   To rate problem severity, Jeffries et al. (1991) provided their raters with the description of usability problems exactly as they had been written by the participants. No attempt was made to disguise the UEM group from which the problem statements came. In fact, the authors told us that problems identified by user testing were rated as more severe than other problems and that these ratings "may reflect a bias on the part of the raters" because "it was easy to tell which problems came from the usability test" (p. 122). If severity had to be judged by ratings (as opposed to a more objective method), then the raters should not have known (i.e., been blind to) which UEM group found what problem.

Construct Validity

Causal.   Because the meaning of the term *heuristic evaluation* has changed substantially since Jeffries et al. (1991) wrote their report, researchers have misinterpreted what this study has to say about UEMs. Heuristic evaluation has gone from being a primarily descriptive term to referring to a well-defined technique for evaluating usability. This problem was noted by Virzi, Sorce, and Herbert (1993) and was confirmed by Jeffries in an e-mail exchange (R. Jeffries, personal communication, May 18, 1995). Jeffries et al.'s use of this term is faultless; indeed, it is exemplary as their description of their methods was such that a careful reader, such as Virzi, could map Jeffries et al.'s methods onto the changing UEM labels. However, as a reader, Virzi's diligence is the exception, not the rule. For example, a clear misreading is Nielsen's (1994b) citation of this study as showing that "independent research has indeed confirmed that Heuristic Evaluation is a very efficient usability engineering method" (p. 32). Such misreadings have lead to miscommunications and false conclusions by researchers and practitioners alike.

Effect: How Problems Were Counted. Usability problems were counted for each UEM as they were found by participants in that UEM condition. There was no attempt to classify similar problems across UEMs.[7] Therefore, we do not know how much, if any, overlap there was in the problems or types of problems found by the different UEMs. Thus, if we do not know whether the problems found using Cognitive Walkthrough and guidelines were actually the same kinds of problems, the conclusion that "the [C]ognitive [W]alkthrough technique was roughly comparable to guidelines" (pp. 123–124) may be misleading. Likewise, if we do not know the extent to which problems found using the different UEMs were unique, the conclusion that "the [H]euristic [E]valuation technique as applied here produced the best results" (p. 123) is not as informative as it might otherwise be.

External Validity

A very particular combination of settings, evaluators, and UEMs was used in this study. This combination would make it impossible to generalize the results to other persons, places, or variations on UEM operations.

Conclusion Validity

As discussed in our overview of Jeffries et al., these authors cite as one of their goals the determination of how expertise affects performance. They fol-

---

7. For a given UEM, duplicate problems were identified by three external (i.e., nonparticipant) raters and eliminated. Our argument here applies to the comparison of problems found by one condition to problems found by another.

low up on this in their conclusions, stating that Heuristic Evaluation is dependent on "having access to several people with the knowledge and experience necessary to apply the technique" (p. 123). However, the study was not designed to examine either the number of people necessary for an Heuristic Evaluation or the knowledge and experience required for conducting the evaluation. Thus, this conclusion goes beyond the data.

Summary of the Review of Jeffries et al. (1991)

If Jeffries et al. (1991) had been cast as a case study (and appropriate changes made throughout), the paper would have provided a snapshot of the trade-offs facing Hewlett Packard in deciding how to do usability analyses in the late 1980s. Unfortunately, the work was presented as an experimental comparison of four UEMs, and several misleading conclusions were drawn. Those regarding one UEM versus another were weak because of low power and lack of statistics, as well as uncontrolled differences (i.e., setting and selection) among groups. Claims made about the types of problems found by each UEM are problematic for the same reasons. In addition, there are construct validity issues with how problems were counted. Finally, conclusions regarding the strengths of different types of evaluators went beyond the scope of the study. Overall, the design and scope of the study did not support the inferences made regarding cause and effect or generality of the results. (See Figure A-1 for a detailed list of claims and the validity problems that weaken them.)

5.2. Karat, Campbell, and Fiegel (1992)

Overview

Karat et al. (1992) compared user testing with a walkthrough technique that combined scenarios with guidelines (heuristic walkthrough by our definition). Forty-eight participants were drawn from a participant pool consisting of users, developers, and UISs. They were assigned randomly to three conditions: user testing (two groups of 6 individuals), individual walkthrough (two groups of 6 individuals), and team walkthrough (two groups with six teams of 2 individuals per team). One group in each condition evaluated one integrated office system (text, spreadsheet, and graphics), whereas the second group evaluated a second integrated office system. (See Figure 4.) During a 3-hr session, participants used the technique to which they had been assigned to learn about the system, freely explore the system, work through nine scenarios, and complete a questionnaire.

Karat et al. used the outcomes of this study (a) to compare the number of problems found, problem severity, and resources required by each UEM; (b)

Figure 4. **Participants per group in Karat et al. (1992).**

System	User Testing	Individual Walkthrough	Team Walkthrough	Total Per System
System A	6	6	6 teams of 2	18
System B	6	6	6 teams of 2	18
Total per UEM	12	12	12 teams of 2	

Note. UEM = usability evaluation method.

to determine if differences between UEMs generalized across systems; and (c) to examine how the characteristics of walkthroughs influenced effectiveness (e.g., the effectiveness of individual versus team evaluations, the effect of evaluator expertise, the value of scenarios). Although the study's design was fairly strong, a few validity problems undermined the soundness of some of the authors' conclusions.

Statistical Conclusion Validity

Sample Size. Recognizing that, in their team walkthrough condition, the team was the unit of analysis and not the number of individuals per team, Karat et al. went to some lengths to ensure that each group had 6 participants (6 teams or 6 individuals; see Figure 4). Although 6 participants per condition was not a large number, 12 participants for each UEM condition and 18 per system (see Figure 4) may have provided enough power to guard against the Wildcard effect. However, as we discuss in the next section, the authors' analyses failed to take advantage of this strength in the study's design.

Data Analysis. The majority of the comparisons reported followed from the design of the study and clearly were not opportunistic efforts to capitalize on chance factors (see Section 4.1). However, the comparisons made and conclusions drawn rested on simple descriptive statistics (e.g., averages and percentages) or χ^2 analyses that did not take into account the variability of participants. With 18 participants per system (see Figure 4), tests that considered participant variability should have been performed. Without the support of statistics to compare group differences against participant (subject) variability, we should not infer that differences in group performance were due to treatment rather than participants; that is, the Wildcard effect has not been ruled out.

Internal Validity

A possible *general selection* problem exists. Two UISs administered user tests; in contrast, a combination of users, UISs, and SWEs completed the walkthroughs. This difference may have contributed to the finding that user testing was better than walkthroughs.

Construct Validity

Dealing With Mono-Operation and Mono-Method Bias.   Karat et al. directly compared two ways (or sets of operations) of conducting walkthroughs: individual versus team. This comparison was a noteworthy attempt to move from simplistic statements ("UEM-A is good") and offer more useful guidance ("If you have 3 UISs, have them use UEM-A as a team"). The use of two different business software packages in the same experiment was also noteworthy. This practice probably reduced mono-method bias and facilitated the generalizability of their findings. However, their manipulation would have been more useful if the authors had been able to characterize the nature of the differences between the two systems without giving away proprietary information. As is, readers are provided with no guidance in mapping from the particular systems used by Karat et al. to characteristics of systems that they might wish to evaluate.

Confounding.   Karat et al. claimed that "all walkthrough groups favored the use of scenarios over self-guided exploration in identifying usability problems. This evidence supports the use of a set of rich scenarios developed in consultation with end users" (p. 403). Without careful qualification, the design of the study does not permit us to conclude anything about the value of scenarios. First, all participants completed self-guided exploration first and scenarios second. Hence, experience gained during the self-guided phase was available to these groups during the scenario phase. This uncontrolled order effect of self-guided exploration on scenarios threatened the causal construct validity of their claim by confounding the treatments. Second, although participants liked the scenarios, no evidence was presented to suggest that the scenarios helped them find more problems. Thus, this claim goes beyond the scope of the study.

Effect Construct Validity.   To support their conclusion that "team walkthroughs achieved better results than individual walkthroughs in some areas" (p. 403), Karat et al. provided a χ^2 analysis of the number of *problem tokens* found. This analysis showed a statistically significant difference favoring team walkthrough over individual walkthrough. However, Karat et al.

classified problems in four ways: tokens, types, problem areas, and *significant problem areas*; it appears that these other measures were not necessarily consistent with the authors' interpretation. For example, for problem areas, the means for the team and individual walkthrough groups were identical. In terms of significant problem areas, teams found more problems than individuals for System 1 (mean of 3.83/team vs. 3.00/individual) but not for System 2 (mean of 2.33/team vs. 2.83/individual). Hence, the conclusion favoring team walkthroughs over individual walkthroughs depends on the usability measure that was chosen; that is, had the analysis been based on the number of problem areas or significant problem areas, it would have told a different story.

External Validity

The external validity of the findings must be tempered by two considerations. First, the development team members (SWEs) used here were not members of the team that developed the product. Thus the finding that "users and development team members can complete usability walkthroughs with relative success" (p. 403) may not generalize to SWEs that are asked to evaluate software that they have developed.

Second, the authors can generalize to mixed teams of GUI users, UISs, and SWEs; however, readers should be careful about generalizing these results across the population (see the discussion of External Validity in Section 4.2) to teams composed of just graphical user interface (GUI) users, just developers, or just UISs. For example, "relative success" may be due to the synergy of a mixed team, solely due to the SWEs, or solely due to the GUI users.

Conclusion Validity

Unfortunately, many of authors' conclusions are either beyond the scope of their study or contradicted by their own data. For example, the authors claimed that studies by Jeffries et al. (1991) and Desurvire et al. (1991), as well as their own study, "provide strong support for the value of [user interface] expertise" (p. 403). This claim goes well beyond the data. User interface expertise was not varied in Jeffries et al. (1991), and it was not varied here.[8] Because user interface expertise was not manipulated in this study, this conclusion is beyond the scope of the study. In a discussion of the significant problem areas

---

8. In addition to Jeffries et al., the authors cited Desurvire, Lawrence, and Atwood (1991) as support. This work is a two-page SIGCHI Bulletin report of a conference poster session. It is impossible for us to drawn conclusions based on the description given in this source.

identified across methods, Karat et al. claimed that "these methods [user testing and walkthroughs] are complementary and yield different results; they act as different types of sieves in identifying usability problems" (p. 403). This is a contradicted conclusion. Based on the data presented in Karat et al.'s Table 4, the UEMs did not yield different types of data. Across the two systems, the user testing group identified 21 problems (called significant problem areas by the authors) not found by the team or individual walkthrough groups. The two walkthrough conditions together found only 3 problems not found by the user testing group. Thus, walkthroughs did not catch many problems missed by user testing.

Summary of the Review of Karat et al. (1992)

This study handled most of the threats to internal validity well and thereby provides a model of how experimental research can be conducted within a corporate environment. The mixed nature of the groups limits the generalization (external validity) of their findings. Given the evaluation philosophy at IBM when this research was conducted, this limit is both reasonable and fair; however, the authors might have stressed this limit in their Discussion and Conclusion sections more than they did. There also are minor construct validity problems concerning several issues. These problems raise concerns with how the study was interpreted, but none of the problems can be considered a fatal flaw. The main failing of this study was with statistical conclusion validity. Few statistical tests were reported, and those that were reported failed to control for the Wildcard effect. Hence, although the results regarding the superiority of user testing to walkthroughs may be interesting and suggestive, they may not be generalizable beyond this study's testing conditions. In addition, some of the claims made about the study have problems with conclusion validity. Thus, several of the findings should be considered with caution. (See Figure A-2 for a detailed list of claims and validity problems.)

5.3. Nielsen (1992)

Overview

In this paper, Nielsen described (a) a study in which he examined whether the probability of finding usability problems increased with usability expertise as well as domain (voice response systems) expertise and (b) a study in which he classified outcomes for six Heuristic Evaluations of different user interfaces along several dimensions. In the expertise study, Nielsen had three groups complete a Heuristic Evaluation of a printed dialogue (as opposed to a running system). Groups consisted of 31 computer science students (novices) who had completed their first programming course, 19 UISs (single experts) who had "graduate degrees and/or several years of job experience in the usability area" but with no special expertise in voice response systems, and 14 double experts who "had expertise in user interface issues as well as voice response systems." Nielsen interpreted the results of his study to provide advice regarding how expertise affects the types of problems found by Heuristic Evaluation.

Evaluation

In the classification study, Nielsen characterized problems found in six different Heuristic Evaluations along the following dimensions: severity (two levels), heuristic (nine heuristics), location (four locations), and type of system (two types). The results of this classification were presented in two forms: the proportion of problems falling into a category per evaluator (mean proportion) and the total proportion of problems aggregated across evaluators. Using these kinds of classifications, he attempted to describe Heuristic Evaluation's strengths and weaknesses in facilitating problem identification. However, problems with assumptions made about the effect construct as well as the approach used to analyze the data severely weakened the validity of these studies, as well as many of Neilsen's claims.

Statistical Conclusion Validity

Sample Size.   In the expertise study, the large number of participants per group (31 vs. 19 vs. 14) should have mitigated the Wildcard effect and should have warranted statistical tests. Unfortunately, none were reported. Although statistical tests were not conducted, possible differences were noted and discussed as if they were real. For example, the conclusion that "usability specialists with expertise in the specific kind of interface being evaluated [double experts] did much better than regular usability specialists without such expertise [single experts], especially with regard to certain usability problems that were unique to that kind of interface" (p. 380) relies on a comparison of group means.

Data Analysis.   In his analysis of Heuristic Evaluations, threats to statistical conclusion validity are even more severe. Not only was the variability of the data unaccounted for when comparisons were made, but the comparisons were selected from a large body of potential comparisons, substantially increasing the probability that apparent differences were due to chance. To illustrate how this invalidated conclusions based on these data, consider the following claim: "Problems with the lack of clearly marked exits are harder to find than problems violating other heuristics" (p. 380). If we isolate the heuristic comparisons, 306 were possible.[9] Thus, this claim was based on 1 seeming difference among a possible 306 differences.

Overall, Nielsen's Table 2 provided a possible 3,546 comparisons by system and another 374 if the data were aggregated by type of prototype.[10] Picking a few particularly attractive differences out of this sea of potential comparisons, as well as failing to consider the stability of the data (their variability) on which these differences were based, was likely to have capitalized on chance factors.

Internal Validity

Unfortunately, the paper does not provide the details necessary to assess thoroughly either study's internal validity (e.g., general or specific selection threats). For example, in the expertise study, we are not told how long each evaluator spent on the task or whether time on task was limited by the experimenter or up to each evaluator. We are also not told whether the evaluation was completed individually or in groups, or anything else about the methodology or conditions of the study. Nevertheless, an instrumentation problem is apparent in both studies.

Instrumentation. In the expertise study, classification of usability problems into minor problems or major problems rests on "a considered judgment" (p. 379). Whereas such an informal basis may suffice for many situations, it does not support a claim as complex as "Major usability problems have a higher probability than minor problems of being found in a [H]euristic [E]valuation, but about twice as many minor problems are found in absolute numbers" (p. 380). A similar problem was inherent in the classification study, as classifications were made in much the same way. (Note that whereas *what* mea-

---

9. To calculate the number of possible comparisons we use $\left[\dfrac{n \cdot (n-1)}{2} \right] \cdot y$, where n is the number of cells and y is the number of DVs. There are nine heuristics and two types of prototypes giving a total of 18 cells ($n = 18$). For the DVs, both mean and aggregate proportions are reported ($y = 2$). Therefore, 306 comparisons are possible.

10. Comparisons made at the system level are based on 12 cells for severity, 54 cells for heuristic, 24 cells for location, and two DVs. Using $\left[\dfrac{n \cdot (n-1)}{2} \right] \cdot y$, where n is the number of cells and y is the number of DVs, to calculate the number of comparisons, we get a total of 3,546 possible comparisons. Comparisons made by type of prototype are based on 4 cells for severity, 18 cells for heuristic, and 8 cells for location. With the two DVs, this makes a possible 374 comparisons.

sures were made is an effect construct validity issue, *how* these measures were made is an internal validity problem with instrumentation.)

Construct Validity

In the expertise study, if we ignore potential problems with statistical conclusion validity then we might conclude that novices named fewer potential problems than the single experts, who, in turn, named fewer potential problems than double experts. Note that the phrase we use is "named fewer potential problems" and not "found fewer actual problems." The *construct validity of effect* was a major weakness in this study. Heuristic Evaluation was not compared to user testing or to any other DV. Anything that an evaluator named was counted as an actual problem. There is no way to be sure that the named problems (intrinsic features) would have corresponded to real problems (pay-off measures of performance).

Conclusion Validity

If potential problems with statistical conclusion validity are ignored then a modest claim such as *experts named more potential problems than nonexperts* would have been supported by the data. However, the broader claim that "usability specialists were much better than those without usability expertise at finding usability problems by heuristic evaluation" (p. 380) was not. There are two issues here. First, no attempt was made in this study to determine if the *named problems* were *actual problems*. It seems likely that there was more than one false alarm (calling something a problem when it is not) or miss (not finding a problem when there is one). Second, whereas the effect of expertise on an evaluator's ability to name problems was clear, the increment contributed by Heuristic Evaluation was not. How many of the named problems were found by expert judgment? How many more named problems did Heuristic Evaluation contribute? As these questions were not addressed in this study, conclusions about them cannot be drawn. A similar weakness existed in the classification study. It was intended to demonstrate the strengths and weaknesses of Heuristic Evaluation by categorizing problems found through that technique. However, it was never established that the problems being categorized were in fact real usability problems—ones that affected performance—and it was not established that Heuristic Evaluation was responsible for revealing those problems.

Summary of the Review of Nielsen (1992)

The author attempted to tackle some critical issues and was creative in finding ways to address those issues based on limited data. However, some of his answers may be misleading. Assumptions made about the effect construct as well as the approach used to analyze the data severely weakened the validity of Nielsen's conclusions. His conclusions assumed that named problems were actual problems and that differences found by opportunistic comparisons and tested only by eyeball statistics were real. Therefore, we need to be careful in how we interpret his advice on expertise and the strengths and weaknesses of Heuristic Evaluation. (See Figure A-3 for a detailed list of claims and validity problems.)

5.4. Desurvire, Kondziela, and Atwood (1992)

Overview

Desurvire et al. (1992) compared the effectiveness of three types of evaluators (UISs [experts], SWEs, and nonexperts) on two analytic UEMs: Heuristic Evaluation and Cognitive Walkthrough. In addition, all UEM conditions were compared with user testing. The Heuristic Evaluation (heuristic walkthrough in our terminology) and Cognitive Walkthrough groups used "paper flow-charts organized by task" to complete six tasks. The user testing group used a prototype of the interface (H. Desurvire, personal communication, October 20, 1995).

The authors' research goals were to determine the value of expertise and to assess the relative strengths and weaknesses of the analytic UEMs compared to user testing. They drew a number of conclusions relating to each of these goals. Unfortunately, weaknesses in the study's design and the analyses used are cause for questioning the validity of all of these claims.

Statistical Conclusion Validity: Sample Size

The user testing group had 18 participants; the six analytic UEM groups had 3 participants each. These participants were distributed as shown in Figure 5. The small number of participants and their distribution among groups raise concerns with statistical conclusion validity as well as internal validity.

Desurvire et al. (1992) attempted to justify the use of 3 participants per group by citing Nielsen and Molich (1990) and Nielsen (1992), who recommended using 3 evaluators when conducting a Heuristic Evaluation. Unfortunately, as discussed in Section 4.1 under Statistical Conclusion Validity, this confusion of the standards appropriate for the usability laboratory with standards appropriate for experimentation threatened the statistical conclusion validity of the experiment. The use of too few participants in an experiment results in low power, unstable estimates of group performance, and a tendency for the Wildcard effect. In addition, the authors were overly zealous in inter-

Figure 5. Participants per group in Desurvire et al. (1992).

Participant	HE	CW	Total
UIS	3	3	6
Non-experts	3	3	6
SWE	3	3	3[a]

Note. HE = Heuristic Evaluation; CW = Cognitive Walkthrough; UIS = usability interface specialists; SWE = software engineers.
[a]Only 3 SWEs were used. The same 3 participated in both the Heuristic Evaluation and Cognitive Walkthrough condition.

preting almost every difference between two numbers as *real* (we count 57 such claims).

As an example, consider Desurvire et al.'s (1992) claim that "[UISs] in the Heuristic Evaluation condition named almost twice as many problems that caused task failure or were of minor annoyance in the laboratory, than [UISs] in the cognitive condition" (p. 99). Because this claim was based on a comparison of totals for 3 participants per group, this statement is misleading. Errors were classified according to problem severity: minor annoyance, caused error, and caused task failure (see Desurvire et al.'s Table 2). The Heuristic Evaluation group found a grand total of four minor annoyance problems, whereas the Cognitive Walkthrough group found two. Likewise, for caused task failure problems, the Heuristic Evaluation group found five problems and the Cognitive Walkthrough group found three ("almost twice as many"). *These numbers do not represent the average number of errors found per evaluator but the total number of errors found per group.* Although the arithmetic is indisputable—four problems is twice as many as two—the meaningfulness and reliability of these claims are questionable.

Internal Validity: Selection

A specific selection threat exists in that the same 3 SWEs evaluated the same six tasks using both Heuristic Evaluation and Cognitive Walkthrough. Hence, in one of these two conditions (we are not told which condition the SWEs completed first) the SWE group had more experience with the task than the UIS group did, making any comparisons among groups difficult to interpret. The second UEM performed by the SWEs had to have been affected by their increased familiarity with the system, with the six tasks, as well as by their memory of the problems found by the first UEM. Thus, it is hard to assess the validity of the claim that "there were no differences between methods for the [SWEs]."

Construct Validity and External Validity

The aforementioned problem with the internal validity of selection also results in a confounding of treatment conditions. This problem, combined with the previously noted problems with statistical conclusion validity, undermine the possibility of anything but a limited generalization of these findings.

Summary of the Review of Desurvire et al. (1992)

It is evident that the authors were eager to share their research and were careful to identify many important questions. Unfortunately, they failed to recognize the limitations of their study and based many strongly worded conclusions on scant data. The prerequisites for an experimental study—statistical conclusion validity and internal validity—were severely lacking. Due to this lack of prerequisites, we do not offer a detailed evaluation of the study's construct or external validity. We believe that there is nothing that can be safely concluded regarding UEMs or expertise based on this study (see Figure 2 and Figure 3; see Figure A-4 for a detailed list of claims and validity problems).

5.5. Nielsen and Phillips (1993)

Overview

Nielsen and Phillips (1993) compared (a) performance time estimates and (b) the costs of four analytic UEMs and user testing. The four analytic UEMs were Cold, Warm, and Hot heuristic estimates[11]; and keystroke level modeling GOMS[12] (KLM). They asked participants in each of the four analytic UEM conditions to use that UEM to estimate the time it would take users to perform tasks using two interfaces (Dialog Box vs. Pop-Up Menu) to the same database. Each participant in the Cold and KLM groups independently evaluated the systems based on written specifications only. The Warm group based estimates on a prototype of the Dialog Box interface and written specifications of the Pop-Up Menu interface. The Hot group used running prototypes of both

---

11. Note that unlike Heuristic Evaluation, heuristic estimates do not involve the use of a set of guidelines or, as far as we can tell from the text, any other instructions from the experimenters; thus, we consider the Cold, Warm, and Hot heuristic estimates to be a type of expert review (see Figure 1).

12. KLM is the quickest to do, least theoretically motivated, and least accurate of the GOMS family of analysis techniques. Although Nielsen and Phillips (1993) referred to this in their paper as *GOMS analysis* we refer to it here as KLM to avoid confusing this with other types of GOMS analyses (e.g., see John & Kieras, 1996a, 1996b).

systems when producing their estimates. Time estimates obtained from the four analytic UEM groups were compared to actual times for a user test group.

Based on the outcomes of this study, the authors drew conclusions about the relative effectiveness and costs of the UEMs, as well as their reliability. They also made claims about each UEM's ability to support evaluators in making absolute and relative estimates of system usability. However, weaknesses in their methodology and analyses threatened the validity of many of their claims.

Statistical Conclusion Validity

Sample Size.   Groups appear to have been sufficiently large to avoid the Wildcard effect. There were 12 Cold, 10 Warm, 15 Hot, and 19 KLM evaluators, and 20 participants in the user test.

Appropriateness of Statistics.   Formal statistical tests were not used; that is, conclusions in this study were based on an inspection of the means, without considering variability. In fact, the data were highly variable and the authors' own estimates (see their Table 2) show that they would have needed many more participants to establish reliable measures of group performance. For example, the claim that "heuristic estimates were better in the hot condition where estimators had access to running versions of the two interfaces, than in the cold condition based on specifications only" (p. 221) is based an inspection of means with very high variability.

Independence of Measures.   The authors assessed the cost of each UEM by comparing the average time to complete evaluations. Unfortunately, as discussed in the following, the Warm group had prior experience with the system. Recognizing that this prior experience would have distorted the estimates of how long it took the Warm group to perform the task, the authors used the average of the Cold and Hot group evaluation completion times to assign a completion time to the Warm group. This assignment of time to the Warm group violated the statistical conclusion validity assumption of independence of measures.

Internal Validity

Selection.   A *general selection* problem existed in that the expertise and background of the KLM and the heuristic estimation groups were not equivalent. "For all three heuristic estimation conditions, the evaluators were usability specialists with an average of nine years of usability experience" (p. 217). For the KLM condition the evaluators were undergraduates doing their second KLM assignment at Lewis & Clark College. Far from experts, half were psychology majors and the rest were English, art, physics, and history majors (E. Nilsen, personal communication, April 18, 1995). (Note that some of these observations have been made by John, 1994.) Thus, any comparisons between the KLM and heuristic estimations (e.g., "GOMS and heuristic estimates were about equal …, " p. 221) are questionable.

In addition, a *specific selection* threat arose from differences between the heuristic estimation groups in prior experience with the system being evaluated. Warm group UISs had completed a Heuristic Evaluation, rated severity problems, and heard a complete explanation of the full system prior to participating in this study. All in all, they spent about 2½ hr on such activities. The other groups had no prior exposure to the system. The effect of this prior experience on heuristic estimation is unknown and makes any comparison involving the Warm group suspect.

Settings.   Differences in setting were also likely threats to validity. Conducting evaluations in the Bellcore workplace versus in a Lewis & Clark College dorm room must have been very different experiences that may have affected the findings in indeterminable ways. Again, this would have affected the validity of comparisons involving the KLM group.

Instrumentation.   The college students in the KLM group were not consistent in how they estimated the time it took them to do the evaluation. Some of the students included time spent to "write a memo to their fictional manager explaining the approach and their recommendations for the new application" and some did not (E. Nilsen, personal communication, April 18, 1995). This is an instrumentation problem that threatens internal validity.

Construct Validity and External Validity

The problems noted previously with internal validity undermine the possibility of anything but a limited generalization of these findings.

Conclusion Validity

Nielsen and Phillips (1993) claimed that "performance estimates from both heuristic estimation and GOMS analyses are highly variable" (p. 221). This claim does not appear to be supported by their data. Although outcomes for heuristic estimation do appear to be highly variable, outcomes for KLM analyses do not. If the authors had in mind an implicit comparison with user testing, then the variability of KLM estimates (with novice, KLM analysts) compared well with the variability of user testing. From their Table 1, we see that the "Standard Deviations as % of Means" for the cold, warm, and hot heuristic estimate conditions were 108%, 75%, 52%. In contrast, for KLM (shown in the table as "GOMS") and user testing the ratios were 19% and 17% (see John, 1994, for a replication of the KLM results). Although it is clear that the heuristic estimations of experts in all three conditions were highly variable, the estimates from KLM novices were not. Hence, the part of this claim that applies to the KLM estimates is contradicted by the data.

Summary of the Review of Nielsen and Phillips (1993)

The authors used enough participants to estimate variability and provided estimates of this variability in their presentation of the data. In fact, these variability estimates strongly support a point we have been trying to make: Studies of HCI cannot ignore the *Wildcard effect*. Normal, random variability among participants can result in highly variable outcomes, making the comparison of techniques difficult.

Unfortunately, the authors failed to acknowledge the limits that their highly variable results placed on statistical conclusion validity. Several comparisons were made despite unreliable estimates of group performance. In addition, the authors appeared to recognize problems with selection (both general and specific) but failed to qualify their statements in the conclusion section. Consequently, if time-pressed practitioners were to skip the body of this study and simply read the conclusions, they would be misled.

We find the conclusion that analytic UEMs are best used to make relative rather than absolute comparisons to be convincing. However, we disagree with the authors regarding other claims. First, we believe that the Warm group must be ignored for reasons discussed previously. Second, the claim that there are differences in the performance of the Cold versus the Hot heuristic estimation groups is not well supported—especially if one were to consider relative estimates rather than absolute estimates. Third, with the key factors of training and expertise stacked against them, the KLM group did amazingly well.

5.6. Summary of the Reviews

Our disheartening conclusion is that each of these five influential studies had problems in demonstrating cause-and-effect and generality—the raison d'être for using the experimental method (see Section 3). All suffered from important problems with statistical conclusion validity. Low power (too few participants), failure to control for the Wildcard effect (by using the appropriate statistical tests), and the tendency to yield to the allure of reporting numerous "particularly attractive differences" afflicted each of them to varying degrees. However, even if we assume that the differences reported were real, our confidence in the internal validity of these studies is low. With the exception of Karat et al., problems with setting, instrumentation, or selection provide explanations of the results that rival those proposed by the researchers.

Conclusions regarding the generality of the findings are not much brighter. On the positive side, we find little to fault in how individual[13] experimenters handled causal construct validity and one case, Karat et al., in which the issues of mono-operation and mono-method bias were directly addressed. In contrast, effect construct validity was much more problematic; it is not clear that what was being compared across UEMs was their ability to assess usability (we have more to say about this in the next section). However, if, despite all of this, we could accept the reported results at their face value, we would not be able to generalize beyond the specific persons, settings, and times used by the experimenters. The exception, again, is Karat et al., in which we would be willing to generalize to other teams composed of a mixture of UISs, SWEs, and users.

To varying degrees, the researchers engaged in the practice of offering experience-based advice in their summary and conclusions sections. Unfortunately, they did not take the care needed to distinguish such experience-based advice from experiment-based claims.

6. OBSERVATIONS AND RECOMMENDATIONS

Here we first discuss the most important issue facing usability researchers and practitioners alike: the construct of usability itself. We then take a final look at the four types of validity and the tendency to draw conclusions that are either beyond the scope of the study or contradicted by the data. For each problem type we offer recommendations for avoiding the problem and provide examples drawn from lesser cited UEM studies in which such problems have been avoided.

6.1. Predicting and Measuring Usability: Effect Construct Validity

Analytic UEMs examine the intrinsic features of an interface in an attempt to identify those that will affect usability (the payoff) in some way: errors, speed of use, difficulty of learning, and so forth. The desired mapping for ana-

---

13. As stated earlier, we believe that most issues of causal construct validity are issues that must be addressed by the field as a whole but not, necessarily, within the context of a single experiment.

lytic UEMs is from features-to-problems. Empirical UEMs typically begin with payoff measures and attempt to relate these measures to intrinsic features of the interface that can be changed to eliminate the payoff problem. The desired mapping for empirical UEMs is from problems-to-features. Although it has been shown that performance (payoffs) can be linked to design (or intrinsic) features (see, e.g., Franzke, 1995; Gray et al., 1993), this correspondence cannot be assumed and the links must be carefully forged.

Hits, False Alarms, Misses, and Correct Rejections

It is a sure bet that no UEM is perfect; any UEM will detect some problems and miss others. Figure 6 shows a detection table for a hypothetical UEM. If the UEM claims that A and B are problems, and they are: These are hits. If it claims that they are problems, and they are not: These are false alarms. Likewise, if it claims that C and D are not problems, but in truth they are: These are misses. Finally, if it claims they are not problems, and they are not: These are correct rejections.

All four cells are important. When our UEM claims that something is a problem, how confident are we that this claim is a hit rather than a false alarm? Are we confident enough to recommend (or insist) that resources be devoted to fixing this problem? Likewise, when our UEM says that C and D are not problems, but, for example, someone else in the team (e.g., boss, marketer, SWE, etc.) thinks that they are, how confident are we in saying these are not problems? The issue facing the practitioner under these circumstances is far from academic.

Unfortunately, the problem-counting approach (used in all of the previously mentioned studies except Nielsen and Phillips) conflates the naming of potential problems with success. This conflation is equivalent to summing hits plus false alarms (the middle row in Figure 6), while ignoring misses and correct rejections (the bottom row of Figure 6). For a UEM to be considered reliable and valid, we need estimates of how it would fill in all cells in Figure 6. This will not be an easy task.

Figure 6 is misleading in that it implies that we have access to *truth* as the final arbiter of usability problems. Reality is more muddled. Ideally, different UEMs would converge in identifying the same set of problems; in reality the problem sets identified by two different UEMs are only partially overlapping. If UEM-A identifies a problem not found by UEM-B, does this represent a false alarm for UEM-A or a miss for UEM-B? For example, Nielsen (1992) stated that "heuristic evaluation picks up minor usability problems that are often not even seen in actual user testing" (p. 378). In his 1994 chapter (Nielsen, 1994b), he claimed that "seventeen of the 40 core usability problems that had been found by [H]euristic [E]valuation were confirmed by user test" (p. 45).

Figure 6. Usability problem detection.

	Truth	
UEM Claims That . . .	Real Problem Exists	No Problem Exists
A & B are problems	Hit	False alarm
C & D are NOT problems	Miss	Correct rejection

Note. UEM = usability evaluation method.

He argued that problems not found were not false alarms for Heuristic Evaluation but were due to the characteristics of the users who were involved in user testing. "It would therefore be impossible to find these usability problems by user testing with these users, but they are still usability problems" (p. 46).

We have some sympathy for Nielsen's argument, because we believe that empirical UEMs (i.e., most types of user testing) have mapping problems of their own. However, the only research that we are aware of that has attempted to link specific intrinsic features as identified by Heuristic Evaluation to specific payoffs was done by R. W. Bailey and associates. It yielded negative conclusions.

In two independent experiments, R. W. Bailey and associates (R. W. Bailey, Allan, & Raiello, 1992) performed usability tests on variations of the MANTEL system that was originally devised by Molich and Nielsen (1990). They conducted user tests for five interfaces. Five users participated in each of the tests, making five groups of 5 users. The first group used the original MANTEL-prime system (as built from the published specifications); the fifth group used the system as redesigned by Molich and Nielsen to fix all 29 problems identified by Heuristic Evaluation (again built from published specifications; MANTEL-ideal).

Based on their observations of the MANTEL-prime group, the experimenters identified and fixed two usability problems to create MANTEL-2. Group 2 used MANTEL-2 and observations of this group were used to identify and fix two more problems, creating MANTEL-3. Group 3 used MANTEL-3, resulting in the identification and fix of one more problem. MANTEL-4 was used by Group 4.

Experiment 2 was similar to Experiment 1. A GUI MANTEL-prime was built and subjected to Heuristic Evaluation. The 43 problems named by Heuristic Evaluation were used to build MANTEL-ideal. MANTEL-2 incorporated two changes suggested by user testing of MANTEL-prime. MANTEL-3 improved on MANTEL-2 by fixing 2 problems identified by user testing and, similarly, MANTEL-4 fixed 2 problems identified by MANTEL-3. Both studies found an improvement from MANTEL-prime to MANTEL-2 and no improvement between MANTEL-2 and any other prototype, including MANTEL-ideal. (A strength of this report is that Experiment 2 essentially replicates the Experiment 1 findings using a different style of interface. This replication greatly increases the generality and construct validity of the findings.) Our biggest concern with R. W. Bailey et al.'s (1992) study is their use of total task time as the sole measure of performance. There may well be usability problems that Heuristic Evaluation is picking up but that are not reflected by such a gross measure as total task time. However, this study stands alone as an empirical attempt to validate the recommendations of Heuristic Evaluation. As such, its predominately negative conclusions suggests that Heuristic Evaluation may name many more false alarms than hits.

Tokens, Types, and Categories of Usability Problems

The issue of mapping intrinsic features to payoffs and payoffs to intrinsic features is not the only one that threatens the validity of the usability effect construct. Another issue arises from the attempt to classify large numbers of individual problem tokens (e.g., an error caused by a user selecting the wrong menu item) into a small number of problem categories or types (e.g., "Be consistent").

To the naive observer it might seem obvious that the field of HCI would have a set of common categories with which to discuss one of its most basic concepts: usability. We do not. Instead we have a hodgepodge collection of *do-it-yourself* categories and various collections of *rules-of-thumb*. Our survey of recent UEM-comparison studies reveals three types of problem categories: those created in the course of the study by the researchers to account for the data they had collected (Jeffries et al., 1991; Karat et al., 1992; Smilowitz, Darnell, & Benson, 1993; Virzi, 1992), established lists of guidelines or heuristics that exist in the open literature (Desurvire et al., 1992; Desurvire & Thomas, 1993; Nielsen, 1992), and one based on theory (Cuomo & Bowen, 1994).

Developing a common categorization scheme, preferably one grounded in theory, would allow us to compare types of usability problems across different types of software and interfaces. However, although such categories may be a boon to the researcher, they may be of limited utility to the practitioner. For example, John and Mashyna (1997) argued that attempts to categorize usability problems have lead us to overestimate the success of UEMs. In a carefully analyzed case study, John compared the problems found by a Cognitive Walkthrough with those identified by user testing. Cognitive Walkthrough found 18 problems that could have been found by user testing. In contrast, user testing found 37 problems that could have been found by Cognitive

Walkthrough. Out of this set of potential problems, only 2 were the same. There were 16 problems identified by Cognitive Walkthrough that were not observed in user testing and 35 problems observed in user testing that were not predicted by Cognitive Walkthrough. John argued that the practice of reducing problem tokens to problem types or categories may mislead us into believing that different UEMs have a higher level of agreement than they actually do. She further argued that knowing what problem types an interface has is not really useful for developers. Developers need to know the specific problem (e.g., a problem with an item in a particular menu) and not the general one (e.g., "there are menu problems" or "speak the users' language"). John's arguments highlight yet another threat to the construct validity of common measures of usability.

Convergent Measures

Attempts to derive a clear and crisp definition of usability can be aptly compared to attempts to nail a blob of Jell-O to the wall. Rather than attempting to find the one best measure, we advocate approaches that attempt to get at usability by multiple converging measures. Of the research with which we are familiar, two efforts stand out. The first, by G. D. Bailey (1992, 1993), used six DVs to measure the overall usability of various interface designs. The second, by Virzi and associates (Virzi et al., 1993), compared Heuristic Evaluation done by double experts,[14] think-aloud user testing, and performance-based user testing in an attempt to identify individual usability problems.

The techniques used in Virzi's performance-based condition were those advocated by Landauer (1988). Users were asked to complete each task as quickly and accurately as they could, without talking aloud. The actual times taken by these participants were compared with ideal times (the time needed to complete tasks without error). To identify problems the researchers "looked for tasks and subtask times with high variability and for those that took longer than the 'ideal listener' times" (Virzi et al., 1993, p. 311). As shown by Franzke (1994, 1995) this technique seems especially well suited to transform time from a rough, overall measure of performance to a tool that focuses attention on the most problematic aspects of an interface. Such microanalyses of payoff, when used in conjunction with analytic UEMs, should facilitate the attempt to map problems-to-features as well as features-to-problems.

Effect Construct Validity: Recommendations

The issues surrounding the construct validity of effect are vitally important to the success of the UEM enterprise. Elucidating these issues should be a top

---

14. Evaluators who are UISs as well as expert with the domain or application type.

priority of HCI theorists, and deriving reliable and valid means of detecting and classifying usability problems should be a major concern of the HCI research community. Researchers concerned with the effectiveness of analytic UEMs must seek to relate intrinsic attributes to usability payoffs. In validating these relations, we should seek the convergence of multiple performance measures. Because we have no easy way of knowing *truth*, these measures must be carefully and painstakingly analyzed for evidence that various analytic and empirical UEMs do indeed converge on the same set of usability problems. Examples of the types of analyses that we believe will be fruitful in this endeavor are provided by Virzi et al. (1993) and Franzke (1994, 1995). Such research is not quick or easy to do. However, if the HCI research community is to provide HCI practitioners with UEMs that are reliable and valid, as well as quick and easy, then these are costs that we must be prepared to accept.

6.2. Recommendations for Addressing the Four Types of Validity

Statistical Conclusion Validity

The recommendations in this section are simple to state: Most problems with statistical conclusion validity could be avoided by using a larger sample size or using multiple measures from each UIS (or SWE) collected over several sessions.[15] The underlying issue is how to make the most of a limited access to software developers, programmers, and human factors experts at most sites. The studies that had the most problems with statistical conclusion validity tried to test too many conditions at once; however, alternatives exist. Narrowing the scope of the study by reducing the number of UEMs tested is the quickest way to increase sample size and decrease problems with statistical conclusion validity. An example of a well-conducted study with a small sample size is provided by G. D. Bailey (1993). G. D. Bailey's focused study brought multiple measures of usability (as discussed in Section 6.1) to bear on examining one question: Does the usability of interfaces that were designed by programmers differ from those designed by human factors specialists? To answer this question, he collected data from 4 UISs and 4 SWEs over several sessions. Each participant independently built prototypes of the same system. After each declared his or her prototype ready, it was tested with three users. The videotape of each usability test session was provided to the designer with-

---

15. We recommend holding UEM constant but varying the software package being evaluated. This approach has the added advantage of avoiding problems due to mono-method bias.

out comment and without interpretation. Each designer then redesigned and retested the prototype. This cycle continued until each designer called it quits. Between three and five designs were developed by each designer and tested with 3 new user test participants on each iteration. This study yielded comparisons that were statistically reliable and valid with only 4 participants per each of the two conditions (UIS or SWE).

Another solution to the problem of limited access to software developers, programmers, and human factors experts is to use a small number of participants in each experiment but to replicate that experiment with different participants and, perhaps, different software systems. This last strategy has the concomitant effect of increasing our confidence in internal validity, construct validity, and external validity. (Variations of this strategy were implemented by R. W. Bailey et al., 1992, as well as by Karat et al., 1992.)

Ideally, narrowing the scope of the study would result in more focused questions. Rather than broadly asking whether UEM-A is better than UEM-B, the question might become, Does UEM-A find more feedback problems in walk-up and use interfaces than UEM-B? A beneficial outcome of such a focus might be the use of multiple dependent measures (more than one way of measuring the effect construct) and the avoidance of opportunistic unplanned comparisons that capitalize on chance factors. In any event, researchers and practitioners should keep in mind that, if an effect is too unstable for statistics to show a significant difference, then it is too unstable to be relied on as guidance when selecting UEMs.

Internal Validity

With a large enough sample of participants, *selection* problems can be avoided by ensuring that participants from the same participant pool are randomly assigned to conditions. In studies in which differences among participants is the independent variable (e.g., expertise), care must be taken to ensure that all else (e.g., training on the UEM, experience with the software or system being evaluated, etc.) is equivalent.

Instrumentation problems can be avoided by treating the identification, categorization, and severity rating of usability problems with the same experimental rigor called for in other parts of the design. One way to reduce instrumentation problems is to have multiple blind raters (people other than the experimenters) categorize and rate problems. In the ideal case, the raters would not have knowledge of either the conditions or participants. In addition, the order in which problems are rated should be randomized or carefully counterbalanced across raters. Measures of interrater reliability, such as Cohen's Kappa, also should be computed and reported.

Problems with *setting* can be avoided simply by ensuring that all participants in each UEM condition perform the experiment under the same conditions and in the same location. If circumstances do not permit holding conditions and location constant, then care must be taken to ensure that each UEM is tested equally often in each condition–location combination. For example, in an attempt to test 10 UISs on each of two UEMs, we could imagine recruiting 8 UISs from one company, 8 from another, and 4 from a third, with each UEM being tested on-site. However, because the condition–location varies between companies, to be internally valid we would want to use an equal number of UISs from each company in each UEM group.

Causal Construct Validity

Directly comparing different ways of using a single UEM (mono-operation bias) or the effectiveness of a UEM with different types of software (mono-method bias) is more of a concern for the field as a whole than for individual researchers. (However, see R. W. Bailey et al., 1992, and Karat et al., 1992, for examples of how to address these concerns in a single study.) The prime responsibility of individual researchers on these issues is to provide explicit information about the exact operations and methods used.

If only one variant of a UEM is used (e.g., Heuristic Evaluation done by a small group of SWEs), the individual researchers must be explicit about the exact procedures used. Readers must be aware that terminology changes over time. They must carefully check the study's method section to make sure that the way in which the UEM was instantiated and applied accords with their understanding of the UEM.

If only one software package is used, the researchers should try to characterize it in terms of its *use-characteristics*. Ideally, this information would be sufficient to allow other researchers and practitioners to compare their software to that used in the report. For example, knowing that the UEM has been tested using a nonvisual, auditory menu to a voice mail system may cause the researcher or practitioner to be cautious about applying the UEM to a real-time, safety-critical, display-based command and control system. Attempts such as Olson and Moran's (1996) to characterize when, why, and how to use UEMs are a start. However, we need a more thorough classification of use-characteristics. We might model our efforts after Green's (1989), who attempted to discern the underlying cognitive dimensions on which software use may vary. The issue of the cognitive dimensions of software use is a construct validity issue and one in which HCI theorists can serve both HCI researchers and practitioners.

The problem of confounding, or the interaction of different treatments, deserves separate mention. The most highly recommended cure is to use independent groups of participants so that each group is exposed to just one treatment (or UEM). If the same participants are exposed to more than one treatment, then the standard operating procedure of experimental design is to counterbalance. In counterbalancing, each treatment is equally likely to occur first, second, or third and explicit statistical tests can be conducted to determine if treatment type interacts with treatment order (though see, Poulton, 1982, for a discussion of cases in which counterbalancing does not prevent interactions among treatments).

External Validity

External validity is another threat that is more of a concern for the field-as-a-whole than for individual researchers. The standard solution to concerns with external validity is replication. If a finding can be replicated, either by the original experimenters or by a different set of experimenters, then its external validity is bolstered. However, until replication has increased our confidence about the range of evaluators, settings, and conditions to which the results generalize, the responsibility of individual researchers is to explicitly note the possible restrictions to the scope of their findings.

Some might argue that, to affect practice, conclusions must be strongly stated: Otherwise practitioners will interpret the qualifications as doubt and ignore some very fruitful techniques. We argue that if a study leads a practitioner to make a false generalization, and if this generalization has negative consequences, then the credibility of the research enterprise (and the recommended UEM) are damaged severely.

Conclusion Validity

Many of the studies reviewed showed a tendency to go beyond their data in offering advice about how to do usability studies. Indeed, this practice was stoutly defended by a reviewer of an earlier version of this article. Although we can think of no defense for introducing *contradicted conclusions* (conclusions not supported by the results of the study) into a paper, we do believe that the intentions of those who *went beyond their data* (making claims not investigated in the study) to offer advice are good. Their added advice represents attempts to share nonexperimentally acquired experience and expertise with practitioners. Unfortunately, most of the researchers who have done this have not clearly and explicitly separated their experiment-based claims from their experience-based advice. A notable exception to this tendency is provided by Virzi et al. (1993). Although they do not hesitate to offer advice, their discussion of the issues carefully separates experience-based advice from experiment-based claims. We can only reiterate our earlier statement: Unless such care is taken, the advice may be understood as *research findings* rather than the *researcher opinion* that it is.

6.3. Summary of Observations and Recommendations

Studies of UEMs suffer from all four types of Cook and Campbell (1979) validity. The good news is that none of the problems we found are unique to HCI and all can be overcome, or at least mitigated, by following standard, behavioral science conventions for experimental design and analysis.

Problems with the two types of cause–effect validity (statistical conclusion validity and internal validity) can be resolved by individual researchers paying more attention to methodological and statistical concerns. Likewise, many concerns with generalization (construct validity and external validity) as well as the tendency to go beyond the data in reporting conclusions can be handled by following well-known experimental design considerations and reporting conventions.

We dwelt on problems with the construct validity of effect—that is, ways of measuring usability. The studies we reviewed emphasized the problem-count approach to usability with the goal (implicit or explicit) of providing focused feedback to software designers on specific problems that if fixed would increase usability. Unfortunately, by ignoring threats to the effect construct, the message that these studies convey is an erroneous one—namely, the UEM that names the most potential problems is the most effective. If practitioners are to use such quick and easy measures with confidence, then links between interface features and performance outcomes must be carefully forged.

7. CONCLUSIONS

The multitude of empirical methodologies is a strength of the HCI research community, and one of our most potent methodologies is the experimental method. With Cook and Campbell (1979) we believe that "the unique purpose of experiments is to provide stronger tests of *causal* hypotheses than is permitted by other forms of research" (p. 83). Our review has unearthed no inherent obstacle to applying the experimental method to HCI topics. We draw two broad conclusions from our review.

Our first conclusion concerns the two forms of cause–effect validity: statistical conclusion validity and internal validity. The papers we reviewed have adopted methods and statistical tests that are inadequate to demonstrate cause and effect. Some might argue that this demonstrates the difficulty of doing well-controlled research in an applied setting. Although such research might be difficult to conduct, several less influential studies (e.g., G. D. Bailey, 1993; R. W. Bailey et al., 1992; Smilowitz et al., 1993; Virzi et al., 1993) avoid such failings, demonstrating that it is not impossible. The scope of these less influential studies seems to be smaller than the scope of the five studies we reviewed; unfortunately, the broader scope may be the basis for the appeal of the latter.

It has been suggested to us that having some information is preferable to having no information even when that information is bad. This is an argument that we simply do not understand. If either of us were still practitioners and were still responsible for convincing others what changes to make and what things to leave alone, we are convinced that we would rather rely on our own expert judgment than to base decisions on information that we knew to be bad.

Our second conclusion is a twofold one concerning *generality*, primarily the construct validity of effect. First, in most of the studies we reviewed, usability was treated as a monolithic, atheoretical construct. However, usability pays off by increasing performance on one or more *outcomes of interest*. These outcomes of interest vary depending on the people and the task. For example, the outcomes of interest for safety-critical systems are very different than those for ATMs or video games. Second, the name of the game for analytic UEMs is to predict usability payoffs (e.g., time on task) from an examination of intrinsic features (e.g., menu organization). However, there is not a one-to-one correspondence between the two. Indeed, correspondence cannot be assumed but is a theoretical question to be resolved by empirical methods.

An interest in the design of interfaces has been a persistent HCI topic; an interest in the design of experiments has not. In this review, we have attempted to show the importance of experimental design for the HCI community. Small problems in how an experiment was designed and conducted have been shown to have large effects on what we could legitimately conclude from its outcomes. If the outcomes of these experiments were trivial, then such small problems could be safely ignored. We believe that these outcomes are important; it is important to know the relation between intrinsic features and performance payoffs; it is important to know about a UEM's tendency to name hits versus false alarms, to declare that a feature does not present a problem (correct rejection) versus missing features that do present problems; and it is important to know what types of UEMs work best in evaluating which types of software systems. The experimental method is a potent vehicle that can be brought to bear to address these and other core HCI issues. However, to obtain these desired payoffs, we must pay close attention to intrinsic features of experimental design.

NOTES

Acknowledgments. We give our thanks to Deborah A. Boehm-Davis, William M. Newman, Bonnie E. John, Robin Jeffries, Gary M. Olson, Peter G. Polson, and several anonymous reviewers for their comments on earlier versions of this article. We likewise thank the researchers of the studies we reviewed. Everyone we contacted provided complete and timely responses to our many queries. We are heartened by the cooperation provided and believe that it bodes well for the future of HCI research and practice.

Support. The writing of this review was supported in part by a grant from ONR (N00014-95-1-0175) as well as a Fellowship from the Krasnow Institute for Advanced Studies.

Authors' Present Addresses. Wayne D. Gray, Human Factors and Applied Cognitive Program, George Mason University, msn 3f5, Fairfax, VA 22030, USA. E-mail: gray@gmu.edu. Marilyn C. Salzman, U S WEST Advanced Technologies, 4001 Discovery Drive, Suite 370, Boulder, CO 80303, USA. E-mail: mcsalzm@advtech.uswest.com.

HCI Editorial Record. First manuscript received May 10, 1996. Revisions received March 22, 1997, and August 11, 1997. Accepted by Gary Olson and Thomas Moran. Final version received February 25, 1998. — *Editor.*

REFERENCES

Bailey, G. D. (1992). *Iterative methodology and designer training in human–computer interface design.* Unpublished doctoral dissertation, New Mexico State University, Las Cruces, New Mexico.

Bailey, G. D. (1993). Iterative methodology and designer training in human–computer interface design. *Proceedings of the ACM INTERCHI'93 Conference on Human Factors in Computing Systems,* 198–205. New York: ACM.

Bailey, R. W., Allan, R. W., & Raiello, P. (1992). Usability testing vs. heuristic evaluation: A head-to-head comparison. *Proceedings of the Human Factors Society 36th Annual Meeting,* 409–413. Santa Monica, CA: Human Factors Society.

Bovair, S., Kieras, D. E., & Polson, P. G. (1990). The acquisition and performance of text-editing skill: A cognitive complexity analysis. *Human–Computer Interaction, 5,* 1–48.

Card, S. K., Moran, T. P., & Newell, A. (1983). *The psychology of human–computer interaction.* Hillsdale, NJ: Lawrence Erlbaum Associates, Inc.

Cook, T. D., & Campbell, D. T. (1979). *Quasi-experimentation: Design and analysis issues for field settings.* Chicago: Rand McNally.

Cuomo, D. L., & Bowen, C. B. (1994). Understanding usability issues addressed by three user-system interface evaluation techniques. *Interacting With Computers, 6,* 86–108.

Desurvire, H. W., Kondziela, J. M., & Atwood, M. E. (1992). What is gained and lost when using evaluation methods other than empirical testing. *Proceedings of the*

HCI'92 Conference on People and Computers VII, 89–102. New York: Cambridge University Press.

Desurvire, H., Lawrence, D., & Atwood, M. (1991). Empiricism versus judgement: Comparing user interface evaluation methods on a new telephone-based interface. *SIGCHI Bulletin, 23*(4), 58–59.

Desurvire, H., & Thomas, J. C. (1993). Enhancing the performance of interface evaluators using non-empirical usability methods. *Proceedings of the Human Factors and Ergonomics Society 37th Annual Meeting,* 1132–1136. Santa Monica, CA: Human Factors and Ergonomics Society.

Franzke, M. (1994). *Exploration and experienced performance with display-based systems* (Ph.D. Dissertation ICS Tech. Rep. 94–07). Boulder: University of Colorado.

Franzke, M. (1995). Turning research into practice: Characteristics of display-based interaction. *Proceedings of the ACM CHI'95 Conference on Human Factors in Computing Systems,* 421–428. New York: ACM.

Gray, W. D., John, B. E., & Atwood, M. E. (1993). Project Ernestine: Validating a GOMS analysis for predicting and explaining real-world performance. *Human–Computer Interaction, 8,* 237–309.

Green, T. R. G. (1989). Cognitive dimensions of notations. *Proceedings of the HCI'89 Conference on People and Computers V,* 443–460. Cambridge, England: Cambridge University Press.

Jeffries, R., Miller, J. R., Wharton, C., & Uyeda, K. M. (1991). User interface evaluation in the real world: A comparison of four techniques. *Proceedings of the ACM CHI'91 Conference on Human Factors in Computing Systems,* 119–124. New York: ACM.

John, B. E. (1994). Toward a deeper comparison of methods: A reaction to Nielsen & Phillips and new data. *Proceedings of the ACM CHI'94 Conference on Human Factors in Computing Systems,* 285–286. New York: ACM.

John, B. E., & Kieras, D. E. (1996a). The GOMS family of user interface analysis techniques: Comparison and contrast. *ACM Transactions on Computer–Human Interaction, 3,* 320–351.

John, B. E., & Kieras, D. E. (1996b). Using GOMS for user interface design and evaluation: Which technique? *ACM Transactions on Computer–Human Interaction, 3,* 287–319.

John, B. E., & Mashyna, M. M. (1997). Evaluating a multimedia authoring tool with Cognitive Walkthrough and think-aloud user studies. *Journal of the American Society of Information Science, 48,* 1004–1022.

Karat, C. -M., Campbell, R., & Fiegel, T. (1992). Comparison of empirical testing and walkthrough methods in user interface evaluation. *Proceedings of the ACM CHI'92 Conference on Human Factors in Computing Systems,* 397–404. New York: ACM.

Keppel, G., & Saufley, W. H., Jr. (1980). *Introduction to design and analysis.* San Francisco: Freeman.

Kieras, D. (1997). A guide to GOMS model usability evaluation using NGOMSL. In M. Helander, T. K. Landauer, & P. Prabhu (Eds.), *Handbook of human–computer interaction* (2nd ed., pp. 733–766). New York: Elsevier.

Landauer, T. K. (1988). Research methods in human–computer interaction. In M. Helander (Ed.), *The handbook of human–computer interaction* (pp. 905–928). New York: Elsevier.

Lewis, C., & Polson, P. G. (1992). *Cognitive walkthroughs: A method for theory-based evaluation of user interfaces.* Paper presented at the Tutorial presented at the CHI'92 Conference on Human Factors in Computing Systems, Monterey, CA.

Lewis, J. R. (1994). Sample sizes for usability studies: Additional considerations. *Human Factors, 36,* 368–378.

Molich, R., & Nielsen, J. (1990). Improving a human–computer dialogue. *Communications of the ACM, 33,* 338–348.

Nielsen, J. (1992). Finding usability problems through heuristic evaluation. *Proceedings of the ACM CHI'92 Conference on Human Factors in Computing Systems,* 373–380. New York: ACM.

Nielsen, J. (1993). *Usability engineering.* Boston: Academic.

Nielsen, J. (1994a). Enhancing the explanatory power of usability heuristics. *Proceedings of the ACM CHI'94 Conference on Human Factors in Computing Systems,* 152–158. New York: ACM.

Nielsen, J. (1994b). Heuristic evaluation. In J. Nielsen & R. L. Mack (Eds.), *Usability inspection methods* (pp. 25–62). New York: Wiley.

Nielsen, J., & Molich, R. (1990). Heuristic evaluation of user interfaces. *Proceedings of the ACM CHI'90 Conference on Human Factors in Computing Systems,* 249–256. New York: ACM.

Nielsen, J., & Phillips, V. L. (1993). Estimating the relative usability of two interfaces: Heuristic, formal, and empirical methods compared. *Proceedings of the ACM INTERCHI'93 Conference on Human Factors in Computing Systems,* 214–221. New York: ACM.

Olson, J. S., & Moran, T. P. (1996). Mapping the method muddle: Guidance in using methods for user interface design. In M. Rudisill, C. Lewis, P. G. Polson, & T. D. McKay (Eds.), *Human–computer interface designs: Success stories, emerging methods, and real world context* (pp. 269–300). San Francisco: Kaufmann.

Polson, P. G., Lewis, C., Rieman, J., & Wharton, C. (1992). Cognitive walkthroughs: A method for theory-based evaluation of user interfaces. *International Journal of Man–Machine Studies, 36,* 741–773.

Poulton, E. C. (1982). Influential companions: Effects of one strategy on another in the within-subjects designs of cognitive psychology. *Psychological Bulletin, 91,* 673–690.

Scriven, M. (1977). The methodology of evaluation. In A. A. Bellack & H. M. Kliebard (Eds.), *Curriculum and evaluation* (pp. 334–371). Berkeley, CA: McCutchan.

Smilowitz, E. D., Darnell, M. J., & Benson, A. E. (1993). Are we overlooking some usability testing methods? A comparison of lab, beta, and forum tests. *Proceedings of the Human Factors and Ergonomics Society 37th Annual Meeting,* 300–303. Santa Monica, CA: Human Factors and Ergonomics Society.

Smith, S. L., & Mosier, J. N. (1986). Guidelines for designing user interface software (ESD-TR-86-278). Bedford, MA: MITRE Corporation.

Virzi, R. A. (1992). Refining the test phase of usability evaluation: How many subjects is enough? *Human Factors, 34,* 457–468.

Virzi, R. A., Sorce, J. F., & Herbert, L. B. (1993). A comparison of three usability evaluation methods: Heuristic, think-aloud, and performance testing. *Proceedings*

of the Human Factors and Ergonomics Society 37th Annual Meeting, 309–313. Santa
Monica, CA: Human Factors and Ergonomics Society.

Wharton, C., Bradford, J., Jeffries, R., & Franzke, M. (1992). Applying cognitive
walkthroughs to more complex user interfaces: Experiences, issues, and recommendations. *Proceedings of the ACM CHI'92 Conference on Human Factors in Computing Systems,* 381–388. New York: ACM.

Wharton, C., Rieman, J., Lewis, C., & Polson, P. (1994). The cognitive walkthrough
method: A practitioner's guide. In J. Nielsen & R. L. Mack (Eds.), *Usability inspection methods.* New York: Wiley.

Figure A-1. Validity of claims made by Jeffries et al. (1991).

Claim	Validity Problems				
	Statistical Conclusion Validity	Internal Validity	Construct Validity	External Validity	Conclusion Validity
"Overall, the [H]euristic [E]valuation technique as applied here produced the best results. It found the most problems, including more of the most serious ones, than did any other technique, and at the lowest cost" (p. 123).	Low power No statistical tests No. of comparisons	Selection Instrument Setting		Sample	
Heuristic Evaluation is dependent on "having access to *several* people with the knowledge and experience necessary to apply the technique" (p. 123).	Low power No statistical tests				Beyond the scope
"Another limitation of [H]euristic [E]valuation is the large number of specific, one-time, and low-priority problems found and reported" (p. 123).		Instrument	Cause Effect		
"Usability testing did a good job of finding serious problems ... and was very good at finding recurring and general problems, and at avoiding low-priority problems" (p. 123).		Instrument	Cause Effect		

(Continued)

Figure A-1. Validity of claims made by Jeffries et al. (1991). (*Continued*)

		Validity Problems				
Claim	Statistical Concluson Validity	Internal Validity	Construct Validity	External Validity	Conclusion Validity
User testing "was the most expensive of the four techniques. ... Despite this cost, there were many serious problems that it failed to find" (p. 123).	Low power No statistical tests	Selection Instrument Setting	Cause Effect	Sample	
"The guidelines evaluation was the best of the four techniques at finding recurring and general problems ... [but] missed a large number of the most severe problems" (p. 123).	Low power No statistical tests	Selection Instrument Setting	Cause Effect	Sample	
"The [C]ognitive [W]alkthrough technique was roughly comparable in performance to guidelines. ... In general, the problems they [Cognitive Walkthrough group] found were less general and less recurring than those found by other techniques" (pp. 123–124).	Low power No statistical tests	Selection Instrument Setting	Cause Effect	Sample Setting	

Figure A-2. Validity of claims made by Karat et al. (1992).

Claim	Validity Problems				
	Statistical Conclusion Validity	Internal Validity	Construct Validity	External Validity	Conclusion Validity
"Findings regarding the relative effectiveness of empirical testing and walkthrough methods were generally replicated across the two GUI systems... the significant differences in the style and presentation of the two GUI systems in the study support the reliability of the results across these types of systems" (p. 402).	Low power No statistical test				
User testing "identified the largest number of problems, and identified a significant number of relatively severe problems that were missed by the walkthrough conditions" (p. 402).	Random heterogeneity				
"Walkthroughs of the type in this study are a good alternative when resources are very limited" (p. 403).			Effect		Beyond the scope
"These methods [user testing and walkthroughs] are complementary and yield different results; they act as different types of sieves in identifying usability problems" (p. 403).			Effect		Contradicted

(*Continued*)

Figure A-2. Validity of claims made by Karat et al. (1992) (Continued).

| | Validity Problems | | | | |
Claim	Statistical Conclusion Validity	Internal Validity	Construct Validity	External Validity	Conclusion Validity
Jeffries et al. (1991), Desurvire et al. (1991), and the current study "provide strong support for the value of [user interface] expertise" (p. 403).					Beyond the scope
"Team walkthroughs achieved better results than individual walkthroughs in some areas" (p. 403).	Low power No statistical test		Effect		
"All walkthrough groups favored the use of scenarios over self-guided exploration in identifying usability problems. This evidence supports the use of a set of rich scenarios developed in consultation with end users" (p. 403).	No statistical test		Confounding		Beyond the scope
"The results also demonstrate that evaluators who have relevant computer experience and represent a sample of end users and development team members can complete usability walkthroughs with relative success" (p. 403).				Sample	

Figure A-3. Validity of claims made by Nielsen et al. (1992).

Claims	Validity Problems				
	Statistical Conclusion Validity	Internal Validity	Construct Validity	External Validity	Conclusion Validity
"Usability specialists were much better than those without usability expertise at finding usability problems by heuristic evaluation" (p. 380).	No statistical test	Unable to evaluate	Effect		Beyond the scope
"Usability specialists with expertise in the specific kind of interface being evaluated [double experts] did much better than regular usability specialists without such expertise [single experts], especially with regard to certain usability problems that were unique to that kind of interface" (p. 380).	No statistical test	Unable to evaluate	Effect		
"Major usability problems have a higher probability than minor problems of being found in a heuristic evaluation, but about twice as many minor problems are found in absolute numbers" (p. 380).	No statistical test	Instrument	Effect		
"Problems with the lack of clearly marked exits are harder to find than problems violating other heuristics" (p. 380).	No. of comparisons No statistical test	Instrument	Effect		
"Usability problems that relate to a missing interface element are harder to find when an interface is evaluated in a paper prototype form" (p. 380).	No. of comparisons No statistical test	Instrument	Effect		

Figure A-4. Validity of claims made by Desurvire et al. (1992).

Claims	Validity Problems				
	Statistical Concluson Validity	Internal Validity	Construct Validity	External Validity	Conclusion Validity
"Heuristic Evaluation is a better method than the Cognitive Walkthrough for predicting specific problems that actually occur in the laboratory, especially for [UIS]" (pp. 98–99).	Low power No statistical test No. of comparisons		Treatment interact		
"[UIS] in the Heuristic Evaluation condition named almost twice as many problems that caused task failure or were of minor annoyance in the laboratory, than [UIS] in the cognitive condition" (p. 99).	Low power No statistical test No. of comparisons		Treatment interact		
"Heuristic Evaluation seems to facilitate the identification of potential problems and improvements that go beyond the scope of the tasks, more so than the Cognitive Walkthrough" (p. 99).					Beyond the scope
"In the Cognitive Walkthrough evaluation, [UIS] are good at predicting time, task and prompt related problems. [SWEs] are good at naming system, time, and prompt related problems. Non-Experts are only good at finding time related problems. In Heuristic Evaluation, [UIS] are good at time, prompt, task, and system related problems. [SWEs] are good at system, time, and prompt problems, and Non-Experts are best at system and keying problems" (p. 99).	Low power No statistical test No. of comparisons	Instrument	Treatment interact		

Quote	Low power, No statistical test, No. of comparisons	Instrument	Treatment interact
"[UIS] focused on problems that violate the heuristic, 'provide feedback', where they are more focused on the user's interaction with the system than the [SWEs] who were ..." (p. 99).	Low power No statistical test No. of comparisons	Instrument	Treatment interact
"[UIS] were the best at predicting laboratory problems that caused task failure, errors, and caused confusion in the users. The [UIS] were better in the heuristic condition than in the Cognitive Walkthrough, and there were no differences between methods for the [SWEs] and the Non-Experts" (p. 99).	Low power No statistical test No. of comparisons		
"[UIS] were also best at predicting the user's attitude as a result of a problem in the laboratory" (p. 99).	Low power No statistical test		
"This study has shown that evaluation methods can identify a number of interface problems, and these methods are particularly useful by [UIS]" (p. 100).	Low power No statistical test No. of comparisons		Treatment interact
"At best, these methods provide only 44% of the problems seen in a laboratory based usability study" (p. 100).	Low power No statistical test No. of comparisons		

Note. UIS = usability interface specialist; SWE = software engineer.

Figure A-5. **Validity of claims made by Nielsen and Phillips et al. (1993).**

Claim	Validity Problems				
	Statistical Conclusion Validity	Internal Validity	Construct Validity	External Validity	Beyond the Data
"User testing still seems to be the best method for arriving at [estimates of user performance], but one should remember that laboratory testing is not always a perfect predictor of field performance" (p. 220).	No statistical tests		Effect		
"User testing was also much more expensive than 'cold' heuristic estimates and somewhat more expensive than [KLM] analyses" (pp. 220–221).		Selection Setting	Effect	Sample	
"Heuristic estimates were better in the hot condition where estimators had access to running versions of the two interfaces, than in the cold condition based on specifications only" (p. 221).	No statistical tests				
"Estimates of the relative advantage of one interface over the other were much better than estimates of the absolute time needed by users to perform various tasks" (p. 221).	No statistical tests				

Quote					
"[KLM] and heuristic estimates were about equal for relative estimates, so heuristic estimates might be recommended based on its lower costs" (p. 221).		Selection Setting Instrument	Effect	Sample	Beyond the scope
"[KLM] analyses were superior for absolute estimates" (p. 221).	No statistical tests	Selection			
"Performance estimates from both heuristic estimation and [KLM] analyses are highly variable" (p. 221).		Selection Setting		Sample	Contradicted

Note. KLM = keystroke level modeling.

HUMAN-COMPUTER INTERACTION, 1998, Volume 13, pp. 263–323
Copyright © 1998, Lawrence Erlbaum Associates, Inc.

Commentary on "Damaged Merchandise?"

Edited by

Gary M. Olson
University of Michigan

Thomas P. Moran
Xerox Palo Alto Research Center

INTRODUCTION

An interesting dialogue about the issues raised by Gray and Salzman surfaced in the first round of reviews, and it occurred to us that this dialogue could be of value to the readers of *Human–Computer Interaction*. We invited a number of people in the field to submit commentaries on the article. These were reviewed and revised, with all commentators seeing each others' comments, as well as Gray and Salzman's final article.

The commentaries span a wide range of perspectives about research methods, theory, and practice. Briefly:

Commentaries 1 & 2. The commentators include some of the authors of the earlier papers that Gray and Salzman criticize. John Karat and Robin Jeffries were both working in industrial settings when they conducted their studies. Karat's commentary (Section 1) gives a bit of the historical context of studies in the field of HCI and defends the usefulness of the studies in that context. Jeffries and Miller (Section 2) explain how they were dealing with the real-world practicalities of producing what data they could versus doing nothing because of lack of idealized experimental controls.

CONTENTS

Commentaries 3 & 4. The next two commentators are both managers of practicing design organizations within major industrial corporations. Lund (Section 3) points out the importance and need for such studies in industrial settings. Industry does not need perfect answers but recommendations to help improve current practices. Practitioners will quickly discover if a recommendation is not useful. McClelland (Section 4) again emphasizes the different priorities of design practitioners. He discusses the issues of the various aspects of the concept of usability, the context of use, the importance of design representations, and supporting the design process.

Commentaries 5 to 8. Several experts in empirical methods served as commentators. John (Section 5) argues that Gray and Salzman's emphasis on experimental evaluation methods is too strong and recommends case study methods as a useful complementary approach to evaluation. Monk (Section 6) steps back and looks at the role and purposes of experimentation. He describes the practices of experimentation in the field of psychology, in which the questions can be narrowly focused. He then articulates the reasons why this practice does not work on the broad questions posed in a field like HCI and especially on a complex issue such as usability evaluation methods (UEMs). Oviatt (Section 7) presents an opposing view defending experimental methods. She argues for a systematic program of large-scale, longitudinal, replicated studies to tackle the complex issues in evaluating UEMs. Carroll (Section 8) also challenges the value of experimentation in being able to get at the most relevant empirical issues.

Commentaries 9 & 10. Finally, there are two commentaries from researcher–designers. Mackay (Section 9) emphasizes the value of a multiplicity of evaluation methods triangulated within and across disciplines to provide a larger overall picture. Newman (Section 10) emphasizes the important notion that evaluation involves the simulation of future situations. For example, even expensive user testing can be misleading if the situations of interest are simulated too narrowly. He recommends that the larger systems context must be taken into account.

---

1. The Fine Art of Comparing Apples and Oranges

John Karat

The author is a Research Staff Member at the IBM T.J. Watson Research Center; his work is currently focused on the improving the design process for interactive systems. Address: IBM T.J. Watson Research Center, 30 Saw Mill River Road, Hawthorne, NY 10532. E-mail: jkarat@us.ibm.com.

As Gray and Salzman point out in their article, the design of interfaces for computer systems is a core topic for many (if not most) of the researchers and practitioners who associate themselves with the field of human–computer interaction (HCI). For this community, it is producing artifacts that meet the needs of a target group that is the primary goal. Methods that can be of assistance in the broad area of system design—whether they are methods for gathering requirements, for turning requirements into design specifications, or for evaluating designs-in-progress—are of interest to the HCI community. As the community produces an increasing number of papers describing methods that claim to provide useful information, it is natural and desirable for independent evaluations of such design methods to also appear. Reports that describe someone's use of a method in a real-world situation are an important evolutionary part of the uptake of methods by the community.

Once a set of methods has been described, inquiring minds want to know how the methods compare on many dimensions (e.g., how much resource it takes to carry them out and how effective they are at what they claim to do). In the early 1990s there was considerable interest in obtaining some comparative information on usability evaluation methods (UEMs). New methods called for learning different approaches, and people wanted to know whether the change would be worth the effort. It was fairly natural for people who had skill in comparing usability of interface elements such as mice, keyboards, commands, menus, and icons to think of studies that compared the various new methods. It was also fairly obvious that such comparisons would be much more difficult to carry out. The desire for information was great, but there were serious questions about how much effort it would take to gather it using an acceptable experimental procedure.

I do not question whether the criticisms Gray and Salzman provide of the work comparing UEMs are technically correct. As an effort to clarify how much we know about the comparative value of UEMs that are as different as those covered in the article, I think Gray and Salzman have done an excellent job. In the HCI community's rush to leave laboratory evaluation of completed designs behind while it seeks out opportunities to more broadly impact early system design decisions, it is useful to have someone remind us of the rules of experimental evidence pointed out in this article. Can we be sure that the UEMs considered in the various studies give valid, reliable information? No, we can't. Do some of the claims made in the UEM papers go beyond the data? Yes, they do. However, I do think that we need to keep in mind the context in which these studies were conducted and to consider what constitutes a "useful publication" for the community in evaluating such evaluations. Even though we were well aware of the flaws in such designs, perhaps we encouraged the authors of such papers to tell us "something" sooner rather than "something more certain" later.

HCI literature is very much a mixed bag. It includes papers that are mostly about artifacts (e.g., about novel applications, interface features, or design tools) along with papers that are mostly about collecting and interpreting data about the use of—and more recently the design of—artifacts. On one hand we have papers that have no data at all—that present artifacts that are judged to be sufficiently interesting to warrant publication. On the other hand we have papers that evaluate and often compare things that range from icon shapes and colors to loosely defined design approaches such as contextual inquiry. The distinction between "data about use of" and "data about design of" is important here. Although at some level we would like the rules for "collecting and interpreting" data to be the same for these two areas—some publication decision makers (e.g., journal editors, conference paper chairs, and submission reviewers) in the HCI field do distinguish between what are "appropriate" collection and analysis techniques in making decisions about the quality of such papers.

I think that there are several factors related to the emerging identity of the HCI field that played a role in shaping the kinds of experiments reported in the UEM studies. First, I believe that at the time of these studies (early 1990s) the HCI field was experiencing a rapid growth in the development of evaluation methods that were tuned to better inform design activities. This was a time in which the field expanded its concept of evaluation and usability, and methods began to be thought of as targeted to informing design rather than simply reporting on achieved usability levels. Second, because these methods were new (and not particularly well documented), reports of their use were seen as valuable to the HCI community. Several of the papers represent initial attempts by practitioners other than the developers of the methods to apply the methods. Third, the ground rules for advancing the field by conducting comparisons for the still-immature methods properly considered the trade-offs between experimental validity, reliability, and rigor in these studies at the time. Although I agree in general with Gray and Salzman that the reviewed papers are a bit weak in separating experiment-based claims from experience-based advice, I think this is partly a reflection of a belief that an imperfect-but-doable experiment was worth doing and reporting on. I buy the criticisms of the tone of the papers, but I still believe they all represent good work worthy of publication. I believe that these studies all are examples of the reasonable compromises that were necessary as the communities with strong empirical traditions (behavioral scientists) and communities with strong technical bases (computer scientists) took it on themselves to seriously try and work together under the HCI umbrella.

Lots of things have happened since the HCI community first began to form. In earlier times (primarily in the 1980s), many more practitioners were interested in experimental design. I attribute this partially to the academic background of many of the early practitioners—in psychology and human factors—and to an emphasis on laboratory evaluation as a primary activity distinct from design. Within my own company (IBM) the distinction between people who designed and people who evaluated things that other people had designed was long intentionally maintained as a way of ensuring impartiality in evaluation. Usability evaluation was aimed at reporting defects that required fixing at a point in time when fixing something would be very costly. You needed to be sure of problems, and experimental design was important to being sure. Even now, when the role of HCI specialists early in design is more accepted, there is considerable discussion about whether the skills needed for design are very different from those needed for evaluation and perhaps should be assigned to different specialists. Without trying to resolve this debate, I think it is safe to say that the methods employed by HCI specialists early in design are quite different—and less well understood—than those employed in evaluation. As the field has evolved, methods for design have become much more interesting to report on to the larger HCI community than results of particular interface evaluations. I would attribute this largely to two factors. First, design-oriented evaluation methods are well understood and thus fare well in the "originality" criteria generally associated with paper publication. Second, straight reports of empirical evaluations have failed to provide the generality needed to make them relevant to the broad community. Perceived relevance—prone to shifts in interests as technology changes—has won favor over timeless rigor.

When we read or review an article in HCI that makes some claim for the merit of something—whether it is an interface widget, interaction technique, or evaluation method—we should always try to ask, How do I know that author's assertions are correct? In the case of the five papers reviewed in the Gray and Salzman article, I have to say that I was not overwhelmed by the evidence provided by any of them, but I did find all of them useful in my own efforts to deal with the newer focus of effectively informing design. Did the authors take some liberties with their claims? Certainly. However, at the time of publication, making specific recommendations that would be useful to practitioners is exactly what was called for, and some information was viewed as better than limiting our publications to conclusions drawn from unreproachable experiments. The design of usable systems is not a well-defined activity.

It would be one thing if the studies reported could have easily accommodated the sort of comments that are raised by Gray and Salzman. I do not believe that this was or is the case. I know many of the authors of the five studies personally, and I am certain that "holes in the design" of the experiments were not brought about by a lack of knowledge of good experimental design (I know that several of the authors have instructed students in experimental design in past lives). Designing an experiment that would compare the value of two or more different methods in informing the design of a system is not the problem; carrying out such an experimental design is the problem. It is easy to imagine what the experiment might look like—multiple design teams designing "the same" system using different methods and evaluating the usability of the resulting systems to compare the methods. For a system of the degree of real-world complexity that would be required to make the results generalizable, it is not easy to imagine who would provide the resources necessary to carry out the experiment or to imagine that the effort would produce results that would justify doing it.

I disagree with Gray and Salzman in their assertion that it is important to the HCI community that we "accept the costs" of doing better experimental work with UEMs. The methods that can be usefully employed as a useful part of design are likely to be quite different for different settings of users, tasks, and context to be targeted by different systems at the implementation level. It is important to recognize this. It is not a temporary phenomenon brought about because of temporarily inadequate theory. It is not brought about because the field of HCI is so young. Methods used in design in much "older" fields continue to change and evolve. The evolution is helped by reasonable attempts to hold up the methods side by side and say something about the way they perform to successfully accomplish the design goal. The incremental gain in knowledge that I imagine would come from the much more expensive studies called for by Gray and Salzman does not seem warranted. The methods are too flexible and their application too tied up in context. Although I could imagine a study to determine whether apples tasted better than oranges, I would not dream of conducting it because I am fairly certain that "it would depend."

Is contextual inquiry better than GOMS analysis? Is fieldwork more appropriate than laboratory studies for designing usable software? These are not really good questions—at least not as posed here. If we want the publications of the HCI field to impact practice, we cannot limit ourselves to papers that pose questions that can be clearly answered by careful experiments. To the best of our current understanding and for the future that I can foresee, design of interactive systems will be an activity in which researchers and practitioners will benefit from having knowledge of and experience with a range of techniques. It is not knowing which method is best that should be the focus of our research, but developing an understanding of when to use what in what proportions. I believe that this calls for considerations that are orthogonal to experimental design issues in deciding what we consider and encourage in our publications.

2. Ivory Towers in the Trenches: Different Perspectives on Usability Evaluations

Robin Jeffries and James R. Miller

The first author is a Distinguished Engineer at Sun Microsystems, where she designs and evaluates user interfaces for Java environments. Address: Sun Microsystems MPK16-304, 901 San Antonio Road, Palo Alto, CA 94303. E-mail: robin.jeffries@eng.sun.com. The second author is exploring consumer applications of Internet technology as part of Miramontes Computing; previously, he was the program manager for Intelligent Systems in Apple's Advanced Technology Group. Address: Miramontes Computing, 617 Stardust Lane, Los Altos, CA 94024. E-mail: jmiller@miramontes.com.

One of the great strengths of the field of HCI is the diversity of our community—the different concerns, perspectives, and techniques we bring to the problems we study. This strength is at the same time our biggest challenge. It is easy to forget the diversity of perspectives that make up the field and to fall back into thinking that everyone approaches a question from the same perspective that you do, or that their interest in an answer is the same as yours. Or, that it should be.

This is what has happened here. In looking back at our original paper (Jeffries, Miller, Wharton, & Uyeda, 1991) and the response of Gray and Salzman, the real issue is that we did not do the study they would have liked us to have done. Because our concerns were quite different from those of Gray and Salzman, it is not surprising that our methods and tactics were different, even to the point of being unsatisfactory in their minds. Hence, to really understand what is going on here, we have to go back to the questions underlying the original study and what we were hoping to achieve by doing it. We do not mean to dismiss the concerns of Gray and Salzman, but we do want to make our differences clear so that the proper contexts exist for interpreting both our work and theirs.

Difference 1: Ecological Validity Versus Precision

Our study grew out of a unique opportunity: We were able to generate or get access to real usability analysis data, collected via four different usability techniques, about a real product. Our intent was to ask broad questions about these techniques because broad questions were best matched to the nature of these data; we knew that we neither had nor could obtain the kind and amount of data that would support detailed statistical analyses. (More on this later.)

Hence, we opted for a high-level analysis of these usability techniques, with the hope of defining a starting point for the comparison of techniques, one that could be refined by future study. At the time this study was carried out, the only published work in this area was Nielsen and Molich's (1990) paper defining heuristic evaluation as a technique and characterizing the cost–benefit trade-offs of different numbers of evaluators. We wanted to give practitioners a basis for choosing among techniques as well as to identify areas worthy of further investigation.

The real question we faced here—indeed, the real point of contention between Gray and Salzman and ourselves—was how to manage the trade-off between methodological rigor and the value of studying a phenomenon in context. Throughout our reading of Gray and Salzman, we were reminded of the passionate debates among experimental psychologists in the 1970s about ecological validity. On one side were the verbal learners, interested in separating out the individual factors that contributed to human cognition. Generally, these researchers studied how people learned associations among words or carried out other abstract cognitive tasks and then abstracted up from there. On the other side were the contextualists, who said that what the verbal learners were studying was so far from any real human experience that it could not possibly tell us anything about how people remember real information.

As is typical in such debates, the ultimate resolution was that both the contextualists and the verbal learners were right; either type of study can be of value, so long as everyone understands what kind of study is being done, why it is being done, and whether it has a chance of answering the questions it poses. We hope that, rather than fighting these same wars all over again, we can learn from the lessons of the past and move on, from both this controversy and the more general question that the HCI community faces of how ethnographic analysis techniques stand with respect to more traditionally controlled experiments. Experimental control and validity in context are both important; some experiments will emphasize one more than the other. Ultimately, however, the field moves forward when we discuss the content of the work, not battle over which approach is right.

In our study, we greatly valued the real-worldness of the data we obtained, even with the limitations that came with them. As card-carrying psychologists, we knew that we would be generally unable to do traditional statistical analyses on the data because of the small numbers of participants or participant teams. Hence, we did the best we could: We documented our participant groups, methods, and results and explained clearly what we did (and did not) find. Yes, we did statistics in those situations in which statistical inferences were legitimate. There are drawbacks to this approach; in particular, it places greater demands on the readers of a study, who must be sure to tie the results and discussion back to exactly what was—and was not—done by the experimenters. Taking sound bites out of the discussion section of any paper is a risky thing to do, but those risks are greater here.

So, one may ask, why didn't we simply get more data, enough so that statistical analyses would be possible? Simply put, we could not—at least, not without compromising the ecological validity of the study. Our study was dependent on access to people who did real software development, functioning in their roles as developers or usability specialists. Using people who were not real developers or usability practitioners as evaluators—say, student programmers as stand-ins—from our perspective, would have done more damage to the study than making inferences from the small amount of realistic data we could gather. It would be nice if the demands of scientific rigor were not in conflict with real-world practicalities. However, as Gray and Salzman repeatedly point out, access to a large enough participant population was a problem for almost all of the studies they discuss; we believe that this is the reality of the situation, rather than evidence of lack of foresight and imagination on our part. Gray and Salzman exhort us to use our limited participant population more effectively. In our study the different evaluators were not interchangeable; the only participants that might have been substituted for one another were the guidelines and cognitive walkthrough groups. However, simply having two cognitive walkthrough groups to compare with one usability test and one set of expert reviewers would hardly have solved the statistical problem, or satisfied Gray and Salzman.

We believe—now, as then—that a better course to understanding a large question such as this is to begin with a broad, general overview of the question, emphasizing the real-world aspects of the problem even at the cost of experimental purity. If successful, such a study will raise smaller, more focused questions that can be addressed through methodologically more rigorous experiments. That is just what has happened here, as a review of the follow-up experiments cited by Gray and Salzman shows. In these studies, some of our findings have been confirmed and others have been challenged, which is fine with us. Of course, we would like to have been right about everything, but this is how science is supposed to work.

Difference 2: Advice to the Practitioner Versus Scientific Truth About Methods

Gray and Salzman set up a view of the "right" way to do HCI experiments that is strongly colored by their standing as academics: They want to advance science. In contrast, the authors of the papers they critique, including ourselves, had the goal of influencing practice. We believe that our field needs both kinds of experiments, and the discipline is not helped by taking a stand that portrays only one of these perspectives as valid.

As we noted previously, the main difference between the two perspectives is the importance of control versus that of validity. In "scientific" experiments, it is important that one be able to isolate the various contributions to the dependent variable. If novices use Technique A and experts use Technique B, then it is impossible to determine whether an observed difference is due to expertise or technique. For exactly the reasons Gray and Salzman describe, causal attribution and generalization require that the contributions to the variable of interest be isolated and varied in a controlled way. Over a series of experiments, researchers will vary technique, expertise, type of software, experience with the technique, and many other factors and eventually develop a description of the factors that contribute to successful evaluation methods. Potentially, this can do a lot to advance the science of HCI—for example, it could lead to the development of new usability evaluation methods (UEMs) or to changes in existing UEMs to make them much more productive. However, manipulations like these are—in the near term—irrelevant to practitioners, who are interested in how the techniques that exist today can be applied to the usability issues they face. Having novices use a methodology that is generally the domain of experts is not of interest. Rather, what practitioners want to know is whether usability engineers using Technique A, in the environment in which Technique A generally appears, will do better than software engineers using Technique B, in the environment in which Technique B generally appears.

In Jeffries et al. (1991) our overarching goal was to make the groups as realistic as possible. Whenever we had a choice about an approach to take, we chose the one that was closest to the way the technique is used or is recommended to be used. From our perspective, this was not a bug in the experimental design, but a feature: By focusing on how things are actually done, we avoided conditions that were not relevant to our question. Of course, those conditions may well be of interest to others, and we encourage their investigation. We did not achieve perfect veridicality—our most obvious compromise being that the software engineer evaluators were not the same software engineers who were developing the system being evaluated. However, many of the aspects of the experiment that Gray and Salzman criticize were intentionally chosen with the goal of realism in mind.

For instance, Gray and Salzman criticize the fact that the expert reviewers did their task in several separate sessions,[1] whereas the guidelines and cognitive walkthrough groups did the evaluation in a single session. All groups were given the same instructions—to fit this evaluation into their other work over a 2-week period. This was consistent with how our evaluators did their work in a

---

1. This is not quite true—some evaluators did, and some did not.

production environment. It is probably not a coincidence that the guidelines and cognitive walkthrough groups worked through their tasks in a single session: Both are group techniques, and the start-up and wind-down costs for a group make it more efficient to work through one long session than to break it into smaller parts. Conversely, the expert reviewers working alone generally found it easier to fit their (less structured) evaluation into chunks of time between meetings or other work. Gray and Salzman believe that because of the confounding of number of sessions with technique we cannot tell whether the observed differences are due to using multiple sessions or to technique differences. We agree; we can and did only make statements about the effectiveness of an entire context, not of an individual factor. However, just as psychologists do not systematically vary or control for phase of the moon in most experiments, because they believe that such effects are small relative to the variable of interest, we did not control for the myriad variables that might contribute to a particular usability evaluation context. We made a choice about the variables that we believed to be relevant and designed our study around them. If Gray and Salzman have alternative theories about the relevance of certain variables—they cite such factors as experimental setting, application domain, and "software style"—we encourage them to design the appropriate study and report the results.

Gray and Salzman arrived at the conclusion that decisions like the ones described previously interfere with our ability to assess the usability of the various techniques. This is simply wrong. The choices we made surely interfered with our ability to assess the individual factors that contributed to the usability of the different techniques, but they in no way biased the conclusions about whether the techniques, in the context they were used, are useful ways to assess usability.

Theories in the Real World

At the end of their article, Gray and Salzman outline some questions of interest in the assessment of usability evaluations and call for more work to be done on the theory of usability. The problem with their foray into theorizing is that it does not connect with the reality it is supposed to model. In their attempt to simplify things (something that all theories do), Gray and Salzman have created a notion of usability that is foreign to those of us who deal with it regularly on a practical level. In their world, usability problems are very black and white things—either something is a problem, or it is not. This premise is at the heart of their attempt to define a signal detection model of usability.

A signal detection analysis depends on two factors: (a) the ability to identify and enumerate the full set of *trials* (in this case, problems that might or might not be found by a particular evaluator or evaluation) and (b) an a priori definition of what constitutes a signal (positive result) and what is noise (negative result). Neither of these conditions is satisfied in the usability domain. First, the full set of trials is not known—for a given application, we only know the usability problems that have been found to date; there is nowhere we can go to obtain the complete set of problems with an application. Second, there is no rigorous way to classify a reported problem as a hit or a false alarm. How do we treat a problem for which all possible solutions break other parts of the design and make some other aspect of the interface worse? Is reporting this problem a hit (because the problem is real, and a solution does exist for it) or a false alarm (because fixing the problem decreases the overall usability of the application)? Conundrums like this have led us and many of our colleagues to question whether there is any absolute notion of a false positive in usability evaluation. If we do not know what the set of trials are, and we cannot reliably determine which trials are the positive ones, a signal detection analysis does not seem very fruitful.

Beyond this, however, Gray and Salzman's attempt to characterize the world of usability problems as black and white ignores the reality of how usability testing and product definition really works. It is easy for a usability engineer to come up with a list of concerns about an interface. However, what counts in real-world usability evaluation—and what usability theory desperately needs to address—is the ability to determine which problems will have enough impact on user productivity to be worth the effort required to fix them, given all the other problems competing for development resources. This notion of problem severity is missing from much of Gray and Salzman's analyses and from the more controlled studies they cite. As an example, we are not surprised by Bailey, Allan, and Raiello's (1992) finding that the correction of many of the interface problems found by a team of heuristic evaluators had little effect on the application's usability. The proper question to ask the heuristic evaluators—we wish Bailey et al. had been able to do so—is not simply "What problems did you find?", but rather "Of the problems you found, which are worth fixing?" Without an answer to this question, we have neither a good understanding of what the evaluators really thought of the problems they identified nor a good emulation of how usability evaluation takes place in the real world.

The ability to make these judgments of cost versus benefit is what distinguishes a skilled usability engineer from a less-skilled engineer; today, this ability is based far more on experience, personal skill, and other intangibles than on the application of theory or even rigorous methodology. This is an important area for future research; we began to address this question in Jeffries et al. (1991), but it is clear that there is much more to do. For it is here that we encounter the essence of usability, in a form that is critical to both theorists and practitioners.

Summing Up

We close by returning to our original point and our fundamental difference with Gray and Salzman: We simply have different concerns, based on our different goals for our studies and our different perspectives on the overall problem of usability and UEMs. We do not want to argue that either approach is more correct than the other, but rather we hope that together we can make progress on the questions that are of common interest to us all. We need to attack both large abstract problems and small concrete ones; we need to emphasize method and theory at the same time we look to the pragmatic relevance of what we do. However, we need to understand, appreciate, and respect the differences in these approaches. Confusing one with the other does neither position any good, and it takes us no closer toward greater knowledge.

References

Bailey, R. W., Allan, R. W., & Raiello, P. (1992). Usability testing vs. heuristic evaluation: A head-to-head comparison. *Proceedings of the Human Factors Society 36th Annual Meeting,* 409–413. Santa Monica, CA: Human Factors Society.

Jeffries, R., Miller, J. R., Wharton, C., & Uyeda, K. M. (1991). User interface evaluation in the real world: A comparison of four techniques. *Proceedings of the ACM CHI'91 Conference on Human Factors in Computing Systems,* 119–124. New York: ACM.

Nielsen, J., & Molich, R. (1990). Heuristic evaluation of user interfaces. *Proceedings of the ACM CHI'90 Conference on Human Factors in Computing Systems,* 249–256. New York: ACM.

---

3. Damaged Merchandise? Comments on Shopping at Outlet Malls

Arnold M. Lund

The author is the senior director of interactive services engineering at U S WEST Advanced Technologies, where he manages groups with design and usability, software engineering, Internet application, and other responsibilities; his research is in metrics for assessing usability. Address: 972 St. Andrews Lane, Louisville, CO 80027. E-mail: alund@acm.org.

Trained as a hard-nosed Midwestern empiricist, I resonate with several aspects of Gray and Salzman's (1997) critique of the empirical comparisons of usability evaluation methods (UEMs) reported by Jeffries, Miller, Wharton, and Uyeda (1991); Karat, Campbell, and Fiegel (1992); Nielsen (1992);

Desurvire, Kondziela, and Atwood (1992); and Nielsen and Phillips (1993)—a critique that could just as easily be applied to many other HCI studies. Their observation that much of the focus of HCI research and practice has been on design and not on building a rich empirical foundation for the discipline is probably not news to most people working in the field. It could even be justifiably argued that for a discipline such as ours (both in its nature and relative youth), the balance may be appropriate. Still, as I read their review and catalog the problems with the studies, I have to ask myself why the studies have been so popular and how the studies passed the screening process to reach the conferences at which they were presented. Apparently the need for guidance is great and there are few alternatives for meeting the need. There must be enough face validity to the studies to satisfy those who use them.

With the methodological problems in the studies, why have practitioners continued to cite and use the studies? After all, one might expect that if the studies lead to erroneous conclusions the negative impact on design would be noticed. There are certainly many opportunities for HCI practitioners to quickly exchange information about what works and does not work, and intense discussions about the value of UEMs and guidelines are the norm within these forums. Although I would not want to defend the methodological weaknesses of the studies, understanding the needs of those who use the studies and the corporate environments in which they work may provide insights into why the studies are influential.

Each of the studies reviewed was conducted in a corporate setting. In some cases the studies were conducted in a research environment within a corporation, but even then the studies tend to reflect corporate sensibilities. For the practitioner, finding methodologies that can improve the usability of design as time and resources shrink is critical. The practitioner's job depends on it. I recently talked with one practitioner whose normal schedule was to define a problem, design and conduct a usability study, and produce a new design within 2 weeks. The practitioner was being asked to reduce his schedule further. In this practitioner's environment, the goal is typically not to produce the perfect study or design, but rather to do whatever it takes to make the design better than it would otherwise be. Corporate HCI research increasingly is aimed at meeting the needs of these practitioners and is conducted under similar pressures.

The absence of comparative academic research in this area is noticeable and distressing. Given that the need for efficient UEMs is high, however, if the research is not available attempts will be made to generate it. The corporate value of the research is based on its face validity rather than its methodological purity. The goal is to find something quickly that may improve the productivity of the usability process. The cost of a false positive is low because a given UEM is typically implemented by someone who can apply expert judgment and who can incorporate alternate methodologies if the situation appears to demand it. If a recommendation turns out not to be useful, it is simply rejected and the search for a better UEM turns elsewhere. It is almost axiomatic, therefore, that much corporate research undertaken in this environment will have limited validity. That is not to say that the work is intrinsically without value, rather its value is found in the specific environment the research is intended to address. Outside of the practitioner's context of application the research is probably more useful for hypothesis generation (when coupled with the researcher's experience) than when an attempt is made to argue that generalizable causality has been demonstrated in the data.

Gray and Salzman point out that if they were still practitioners they would rather rely on their own expert judgment than to base decisions on information that they knew to be bad. Most practitioners would agree, I believe, that they would rather rely on their own judgment than generalizations that they expected to have negative consequences for the usability of a project. Practically speaking, however, the impact of a recommendation of a less than optimal UEM is often low. First, most experienced practitioners are used to combining experience with confounded data to disambiguate the effects of the confounding and draw conclusions. They will accept or reject a comparison of UEMs based on the combination of the study (or studies) and experience. Second, whichever UEM is chosen will be implemented in such a way that the results of its application are also interpreted through that filter of experience. Third, for high priority projects with complex design issues, design and usability professionals often use several approaches to the problem. The ability to integrate effectively past studies, experiences, and current results in the constrained environment of corporate product development is one of the measures of the practitioner's expertise. Many practitioners might even argue that at some level the expertise of the practitioner in interpreting ambiguous data and drawing conclusions that improve usability is more important than any specific methodology used to obtain the data.

It would be interesting to know how the studies reviewed by Gray and Salzman are used by experienced practitioners. Methodologies could be chosen by practitioners based on the recommendations of the studies, or the studies could be used simply as one of many sources of information when deciding on an approach to a design. It would also be interesting to know whether those choosing methodologies based on the studies have reduced the usability of designs, or whether they have just not improved the usability of the designs as much as they might have with different choices. Perhaps Wildcard developers have more of an impact on design than choice of UEM.

One of Gray and Salzman's criticisms concerns the use of statistics in some of the studies. The practitioner's expertise is relevant here as well, in that an element of that expertise is the practitioner's ability to evaluate the relative importance of the results of a study. When I first began working at Bell Laboratories, I recall hearing that if a difference wasn't at least 10% it wasn't worth fighting over with developers. In graduate school, I was used to running at least 20 participants per condition to reach acceptable statistical power for the phenomena we were studying. In industry the 10% difference had to be detected using the 10 or 12 participants that could be paid for in a project budget. In graduate school, outliers or Wildcards were dealt with by using larger samples. In industry, the goal typically is to model the market segment that will most likely purchase the product. Outliers (often defined based on the experienced practitioners expertise) are sometimes either manually removed before analysis or are actually used to identify unique design problems. The smaller samples and the requirement for larger differences if the results are to be meaningful in practice raises the importance of measures of central tendency relative to comparative statistics and, therefore, the familiarity and comfort practitioners have with these techniques. The comfort that many practitioners have with the statistical techniques used in the studies could be another reason why the studies have managed to have an impact.

To summarize, the culture created by the corporate environment is what makes the results of the studies comparing UEMs so popular. There is a great need, and the methodologies employed are not that different from those used effectively by practitioners to make decisions every day. Further, the studies often attempt to recreate plausible product development environments. Although the techniques used to simulate realistic environments introduced confounding that limits statistical conclusion validity, the techniques increase the face validity required if practitioners are going to pay attention to the results. Finally, the cost of an erroneous recommendation is probably low.

How did the studies come to be accepted by the admittedly prestigious conferences in the first place? Why have the authors drawn spurious conclusions that go beyond the data, and why have they not been more ready to acknowledge the limitations of their data? Why have the more limited but better designed studies not had more visibility? The answers may be in the review processes of the conferences. Many conferences attempt to match the interests and expertise of the reviewers with the content of the papers. With relatively limited academic interest in the topic of methodologies that is of such great interest and importance to practitioners, it is reasonable to predict that practitioners would have a heavy influence in evaluating the papers. For many practitioners, an extension of the corporate culture in which they practice is that they are looking for content that appears to provide value, and the content that would be expected to provide the most value would be content that will help them design more effective interfaces and applications, and advice that will help them do it more efficiently. In this scenario, it would not be surprising to see papers making bold pronouncements and that give practical advice rated highly.

Reviewers who are practitioners might very well have problems with the designs. Nevertheless, they would be expected to be more comfortable with the designs of the studies because they would have had to design analogous studies and make similar extensions of the results in corporate projects. They would be prepared to interpret the results based on the filter of their own experience. They would also be more comfortable with trusting the expertise of well-respected names in the field working in leading laboratories because they would have had to depend on the recommendations of experts in the past. They would probably assume that the unreported study that resolved the confounding, for example, was expert experience.

What should be done? Some options are not practical because the environment in which practitioners and many researchers work within corporations is unlikely to change. The need for new UEMs, cost–benefit analyses of their application, and advice on when to use a given UEM will continue to increase. One way to begin to address the problem, however, is for major conferences to reexamine their paper-reviewing procedures and guidelines for rating papers. I believe that the goal should not be to screen more papers out, but rather to create a reinforcement structure that rewards the kind of work we need to advance the discipline. There certainly need to be more tightly controlled empirical studies. We also need studies that produce interesting hypotheses that address real-world problems with explicit qualifications on the limits of the results, and we need richly reasoned recommendations based on a history of professional experience and targeted studies that complement and inform that experience.

The field of HCI, of course, needs to be concerned with the fundamental issue of how to enrich the empirical base required for theory to prosper. As a practitioner and researcher, however, I am particularly concerned with the issue of how to increase the number and quality of UEMs and how to improve their effective application. This seems like it should be an ideal area for academia and industry collaboration. There has been some, but all too little.

As Gray and Salzman point out, we do not even have standard ways to classify and name methodologies. Usability itself remains a fuzzy concept. Although Gray and Salzman generally use it as synonymous with ease of use (and that largely defined in terms of design errors detected), several of us practicing and conducting research in corporate product development organizations have found that there are at least two components to the concept: ease of use and utility. Ease of use and utility influence each other and incorporate additional elements such as fit within the context of application and user expectations about alternatives for achieving goals. Some of us believe that an additional emotional factor (e.g., including elements of pleasure and aesthetic satisfaction) should be added. The area needs structure. It needs a foundation on which new methodologies that address usability more efficiently and that reflect a broader range of perspectives on usability and design can be built. The foundation needs to be empirically based, and it needs to be grown through experiment and theory. It will be out of this foundation that more effective UEMs will be created and the taxonomy of when to apply them will grow.

Industry brings both an opportunity to impact design through research that addresses real product development environments and a rich body of testing artifacts that should result in hypotheses about the information that is gained through specific techniques and their relative utility in a wide variety of product development environments. If funding mechanisms can be created, academic researchers should have the time to conduct the kinds of parametric studies that Gray and Salzman believe should have been conducted to produce more generalizable results. With academia and industry collaborating, both the field and the user should benefit.

References

Desurvire, H. W., Kondziela, J. M., & Atwood, M. E. (1992). What is gained and lost when using evaluation methods other than empirical testing. *Proceedings of the HCI'92 Conference on People and Computers VII,* 89–102. Cambridge, England: Cambridge University Press.

Jeffries, R., Miller, J. R., Wharton, C., & Uyeda, K. M. (1991). User interface evaluation in the real world: A comparison of four techniques. *Proceedings of the ACM CHI'91 Conference on Human Factors in Computing Systems,* 119–124. New York: ACM.

Karat, C.-M., Campbell, R., & Fiegel, T. (1992). Comparison of empirical testing and walkthrough methods in user interface evaluation. *Proceedings of the ACM CHI'92 Conference on Human Factors in Computing Systems,* 397–404. New York: ACM.

Nielsen, J. (1992). Finding usability problems through heuristic evaluation. *Proceedings of the ACM CHI'92 Conference on Human Factors in Computing Systems,* 373–380. New York: ACM.

Nielsen, J., & Phillips, V. L. (1993). Estimating the relative usability of two interfaces: Heuristic, formal, and empirical methods compared. *Proceedings of the ACM INTERCHI'93 Conference on Human Factors in Computing Systems,* 214–221. New York: ACM.

4. Damaged Merchandise: How Might We Fix It?

Ian McClelland

The author is the Manager of User Interface in Philips Consumer Electronics, with responsibilities for activities in research, product development, and process development. His interests include entertainment products for use in domestic and social situations and the development of user-centered design processes to enable the management of product quality. Address: Philips Consumer Electronics, PO Box 80002, 5600 JB Eindhoven, The Netherlands. E-mail: Ian.McClelland@ehv.sv.philips.com.

The Context in Which Practitioners Work

Usability issues are receiving increasing attention as one of the key opportunities for manufacturers of interactive systems to use in distinguishing their products from those of their competition. Current business climates demand that any activity that is part of the product development process must be of demonstrable benefit to justify the time and effort they demand. Activities related to usability are no exception. Usability practitioners have to prove their worth and have to be sure that any disturbance they might bring to the efficient development of products is kept very small, if any at all. So it is increasingly important that practitioners know which methods are best suited to which circumstances.

I am a Human Factors specialist, managing projects that involve practitioners in the design of a wide variety of user interfaces for products and systems, many of which would not be generally described as "computers." In most cases the user interfaces are for use in circumstances that do not conform to the general notion of desk-based commercial or industrial applications, the context for which the concept of usability was originally developed. As practitioners we are partners in a creative process that needs to determine what the appropriate design solutions are for the anticipated usage situation. We are interested in usability methods that can help create as well as evaluate solutions and that enhance the creative process by guiding design decisions based on evidence. Selecting the most appropriate usability evaluation method (UEM) that helps us in this process is what interests us.

Interest in UEMs has grown as evidenced, for example, by the recent issue of *Behaviour and Information Technology* on Usability Evaluation Methods (Scapin & Berns, 1997). This is understandable because, as industry becomes more familiar with managing product development programs with respect to usability issues, the questions of which method is best and how to make effective use of limited resources are asked more frequently. Usability practitioners need to be able to answer such questions.

Practitioners working in product development need UEMs that are effective and efficient. Good UEMs should help to identify usability problems accurately and quickly, and get the problems fixed. Good practitioners are also in limited supply. Consequently, practitioners are always in search of ways to improve their own practices in terms of enhancing their ability to identify the right usability issues and reducing the time and effort required to fix problems. Also, the availability of resources for practitioners to evaluate and improve on their own practices is usually severely limited. Therefore papers that provide useful guidance on which method is best to use are particularly welcome.

Damaged Merchandise?

Gray and Salzman examined five celebrated and, in their own words, influential papers. The strategy adopted by the authors in analyzing the papers in such a systematic way must be applauded. It is a good example for all who want to investigate methods. It is therefore pleasing that the article is being published. It is also reasonably easy to trace back comments on the papers they reviewed in the original papers themselves. The authors were also formerly practitioners and so we see their declared interest in supporting the cause of usability in practice. However, early on the authors make clear that the purpose of their review is not only one of academic interest, but "to convince you that much of what you thought you knew about UEMs is potentially misleading." Furthermore the authors conclude that "there are no generally applicable recommendations that can be drawn from these papers," and there are significant doubts raised about some of the claims made by each paper. As one gets into the article, the authors continue to bring the reader bad news with no good news to compensate. Such conclusions are, to say the least, worrying and discouraging for the practitioner—worrying because of doubts raised concerning the quality of the research published, and discouraging because the review of the papers clearly demonstrates the complexities involved in evaluating UEM in such a rigorous scientific way.

In their review, the authors concentrate on the technicalities of the experimental designs adopted and the claims made by the authors of the studies reviewed. It has to be said that, although the criticisms made are severe, the authors have been, in general, fair and specific. Only one paper, authored by Karat et al., receives significant credit. The recommendations made by the authors are primarily aimed at researchers. In their conclusions the authors outline a number of recommendations in relation to how better studies could be designed to overcome the inadequacies identified in their reviews. For the research community the message is clear: Design better experiments and try to ensure the results can be generalized.

For the practitioner community priorities are different. The studies that the authors reviewed were aimed at developing best practice in terms of improving the design process. This is what interests practitioners. However, the review does not help the practitioner make sensible choices about the methods available. In fact, the conclusions present practitioners with uncertainty about the quality of the research work they have used to date to support their choice of methods. Perhaps this goes too far. What would have been more useful, and what is needed, is for the authors (or other members of the research community) to counterbalance the bad news with good news by summarizing the lessons that can be generalized from their research. (The authors refer to having reviewed 11 papers in all.)

Experience shows that when practitioners are faced with examining a particular design (or design proposal), their choice of method is often based on considering factors that the authors do not seem to have considered in either the review of the studies or the recommendations made. I outline here some of the main factors that need to be considered.

Usability Is Not "One Thing"

The review hardly questions the nature of usability. The reviewers tend to regard usability as one issue. It is clear that this is not the case. Usability is a multidimensional creature that changes its nature according to prevailing circumstances. In other words, the key driver behind usability work is to identify those factors that are, for the user, critical to the successful and satisfactory use of the product and can be interpreted in terms of criteria that lend themselves to systematic examination. Thus we derive a basis for assessing the quality of a design in user–usage terms. What is clear from recent developments both in consumer and professional product areas is that the relevant factors are both very diverse and are changing. For example, the factors of enjoyment and emotional value are becoming potent issues for consumer products. The methods used to evaluate these issues are different from the traditional usability issues of functional performance, task effectiveness, efficiency, and so forth. Comparing methods may be misleading if the methods are inherently suited to different types of usability issues. Various usability issues can be distinguished:

- Direct objective performance: time, errors, and so forth.
- Functional suitability; micro (task effectiveness) and macro levels (job effectiveness).
- Opinions and subjective impressions.

- Learning.
- Safety.

Evaluating user interfaces in terms of objective measures typically requires different approaches from subjective issues of satisfaction, and even more, hedonistic aspects such as *pleasure in use*. One might question whether the more intangible issues come within the usability spectrum. I would argue they do in so far as they are issues perceived by users as relevant to the intended purpose of a design. In this respect the notion of quality in use provides a useful framework into which usability can be positioned as the basis for the measurement of quality.

Consequently, any attempt to evaluate UEMs needs to explicitly map the methods being considered to the criteria the methods claim to evaluate. Correspondingly, the review should have taken this mapping into account in comparing the five studies.

The authors refer to a lack of common framework for the definition of usability. Notwithstanding the previous comments, in Human Factors terms the community does have something of a standard framework, namely, ISO 9241 (Pt. 11). This standard has set the scene with reference to a general set of usability measures, but practice shows that the specific measures used require interpretation and that other issues become important when one moves outside the domain of the ubiquitous "VDT–commercial office" perspective (for which this standard was originally intended).

Availability of Skill or Know-How

Often the availability of a competent individual to set up, execute, and interpret a usability study is a key step. Experience shows that the choice of method is often determined by the professional experience of the person assigned to carry out the work. It is clear that the appropriate skills or know-how required are usually a mix of (a) skills in usability methods and (b) domain knowledge.

In their review Gray and Salzman do not examine the interaction between skill and method. Some reference is made to this question but only in passing. This issue needs to be addressed. A certain level of competence is required to make effective use of methods. Some methods are more heavily dependant on specialist skills than others. Some methods can involve using contributions from complementary professional groups (e.g., domain experts) to support evaluation studies. Some methods will require a high level of technical expertise to execute. In some cases the skills of the participants in studies may also be an important factor. For professional systems, certain levels of domain knowledge and skill may be needed.

Representing the Design

At no point in the review is the form in which the design is represented considered as an issue in relation to method comparison. In practice, the form of representation is a big factor in determining what type of method may be appropriate. It needs to be addressed. In practice, usability evaluation often begins at early phases of development. Practitioners often need to work with incomplete representations of the total interface. On occasion, practitioners do get the opportunity to work with fully working prototypes that perform as they are intended. However, the reasonable observation of many practitioners over the years has been that if usability issues are to be adequately addressed, then evaluation must start early in development. Therefore an important factor in method choice is the form in which the user interface is represented. Does the particular form of representation enable the particular usability issues to be addressed in some reasonable way? Some ways in which design solutions may be represented would be called simulations or prototypes. Often other forms of representation are used to generate feedback for the design team.

The Setting

Gray and Salzman do not discuss the setting used by the studies reviewed as a factor in comparing the studies. Contextual issues have long been identified as a source of significant influence over the usability of systems. When it comes to evaluating design proposals it is important that practitioners give contextual issues due consideration and decide to what extent the design of any evaluation study should take them into account. In recent years more attention has been focused on the issue of whether traditional usability laboratory studies are sufficient in attempting to simulate anticipated circumstances of use for a given design proposal. After all, the implicit assumption (usually) in any usability study is that the study is attempting to predict future performance. Increasingly, concerns about the ecological validity of studies have been raised with due attention to where studies should be carried out, how much intrusive data collection undermines the validity of the data collected, how much demands on participants to do "strange things in strange places" influence their opinions and performance, and so on. Studies of UEMs should take account of how far methods are suited to the different settings in which UEMs might be used. Usability studies may be undertaken in any one of a number of settings: a usability lab, on-site, outside, and so forth. Diverse issues may be known (or suspected) to have a critical influence on user performance: physical, social, organizational, and so forth.

The setting can also be interpreted in relation to the "phases of use." The traditional usability study of a new design proposal often presents a user with a simulation in a usability lab (or similar), and within 1 hr or so, data are generated about, for example, task performance. One can question how much such a circumstance enables the daily experience of use to be evaluated. It could be argued that the scenario sketched here is comparable with the initial learning phase that a user will go through when confronted with a new product. If, on the other hand, the topic of interest was initial perceptions of the user on first encounter with the product (e.g., in a purchase situation), it is clear that quite different methods would be needed to evaluate a design. In short, when reviewing UEMs, account needs to be taken of which phase of use the method targets.

Good Scientific Study Is Required, But ...

If the omissions noted previously were to be addressed in evaluating UEMs using rigorous scientific methods of the sort advocated by the authors, it is clear that valid studies would be very complex. To control for legitimate variables that we know influence decisions in practice, real life needs to be simplified to a point of abstraction that matches the requirements of doing good science. The danger is that the trials become so specific that generalization becomes very hard to do, or the generalization is so general that the practitioner can interpret it in many ways. There is a case for good scientific methods. The community needs good research. However, is it the only way forward? In view of the ambitions and needs of practitioners (as exemplified by the studies reviewed) we, as a community, should be looking into strategies for how we can generate widely applicable recommendations that supports the improvement of practice.

A common idea is the principle of *triangulation*: drawing on experience from different sources to assess, through reinforcement and correspondence with one's own direct experience, whether one is on the right track. In the absence of better guidance, practitioners tend to rely on such an approach (as is common with many professional activities). However, there clearly are risks involved, not least being that prejudices can easily be reinforced and new insights can be missed through a lack of thoroughness. However, we have to make progress, and if "science" cannot provide the answers, professional intuitions and experience have to. The challenge for the community is to use the platforms available to us, in the public domain as well as in our places of work, to more directly address the exchange of experience as a basis for the improvement of practice.

Supporting the Design Process

UEMs are about helping to manage a creative process. For the practitioner a key parameter is the extent to which the UEM chosen can help steer design decisions in a useful way. Attention will often be focused on reducing risks and avoiding design errors, but also enabling more efficient working. UEMs need to be evaluated in terms of their impact on the design process and helping the design process get to an appropriate result. For example, a good method might be one that helps identify the most severe usability problems faced by the design team and helps to identify the appropriate solution. A good method may be one that supports the process of progressive approximation until a point is reached when the risks of making a poor design decision are reduced to an acceptable level. What is good and bad may be the severity of the consequences for the commercial performance of the product.

Conclusions

The Gray and Salzman article has served a very useful purpose, at least for this commentator. It has highlighted again the need for UEMs to be subjected to systematic study. UEM studies need to generate generally applicable guidelines that enable design practice to be improved. Practitioners need reliable advice on the comparative performance of methods not only in terms of identifying usability problems themselves but also in terms of their impact on the design process. Practitioners do not typically have the opportunity to evaluate their methods, but if they are to successfully impact on the core design processes that guide most organizations, knowing which method is best for which circumstances would "add power to their elbow."

Developing good UEMs will continue to be an issue for the foreseeable future as the field develops and broadens in scope. Consideration needs to be given to the breadth of user communities, the breadth of usage situations, and the range of issues that impact on successful product use. Usability itself as a concept is also developing to encompass issues that go beyond the traditional concerns of functional performance—for example, pleasure in use.

So, in conclusion the community needs improved methods. Practitioners need good advice from the research community. The research community needs to understand the problems faced by practitioners.

References

ISO 9241–11, Ergonomic requirements for office work with visual display terminals, Part 11, Guidance on usability. International Organization for Standardization.

Scapin, D. L., & Berns, T. (Eds.). (1997). Usability evaluation methods [special issue]. *Behaviour and Information Technology, 16*(4/5).

5. A Case for Cases

Bonnie E. John

The author is an Associate Professor in the Human–Computer Interaction Institute and departments of Psychology and Computer Science at Carnegie Mellon University; she is currently researching usability evaluation methods, especially engineering models of human performance. Address: School of Computer Science, Carnegie Mellon University, Pittsburgh, PA 15213. E-mail: **Bonnie_John@cmu.edu**.

Gray and Salzman have told us that evaluating usability evaluation methods (UEMs) is hard. In this, I agree. Getting valid results from an experiment comparing engineering processes is even more difficult than getting valid experimental results comparing engineered artifacts. Extreme care must be taken with internal and external validity; the latter cries out for many skilled UEM users examining realistic systems, often over an extended period of time. This endeavor has more in common with medical research (e.g., assessing the effects of different chemotherapy treatments to the long-term recovery of cancer patients) than it does with the typical 1-hr usability test or with traditional psychological laboratory experiments (e.g., the word superiority experiments discussed by Andrew Monk in Section 6).

However, from this common ground, I believe that Gray and Salzman's emphasis on experiments is too strong. There are at least two dangers in Gray and Salzman's unremitting focus on problems with experimental design. First, some readers may conclude that valid but narrowly focused, controlled experiments are guaranteed to provide understanding of UEM use and effectiveness. Second, some readers may conclude that valid, narrowly focused, controlled experiments are the only way to gain that understanding. Either of these conclusions would be false and potentially dangerous to the development of HCI as a discipline. I use the remainder of this commentary to show why.

Fallacy: Narrowly Focused Controlled Experiments *Will* Provide Understanding of UEM Effectiveness

Narrowly focused controlled experiments tend to have simple dependent measures, things you can count. For comparing UEMs, these measures could be number of usability problems found (either absolute number of instances or categorized by types or by severity, etc.), time to do the analysis, accuracy of the predictions as compared to user tests or field reports, and so forth. Gray and Salzman discuss the dangers of effect construct validity in detail (Section

6.1), even citing my own position that specific instances of usability problems are necessary both for comparing UEMs and for fixing interfaces (John & Mashyna, 1997). I will not repeat their arguments here.

However, even if the outcome measures are truly of interest, and a significant difference in outcome measures can be established between UEMs, experiments do not always tell enough about what caused the outcome to be of pragmatic use. For instance, in Gray's own work (Project Ernestine, with me and Mike Atwood), he discussed a reaction to the statistically significant and important results of a well-designed field trial comparing a new workstation to an old workstation: "The data from the trial were so counterintuitive that, in the absence of a compelling explanation as to why the proposed workstation was slower than the current one, the tendency was to blame the trial instead of the workstation ... " (Gray, John, & Atwood, 1993, p. 294).

In that study we were comparing two workstations that performed exactly the same task, with precisely the same straightforward measure of interest (time to perform a unit task). How much more counterintuitive, or at least difficult to interpret, will be the results from experiments comparing processes that are designed to be performed by different numbers of people, with different backgrounds (e.g., software developers, end users, or usability specialists), using different levels of prototyping (e.g., prose and pictorial specifications or running prototypes), and making different time investments, especially when they give different outcomes like quantitative predictions of performance time or qualitative predictions of potential user problems? My belief is that "without a compelling explanation as to why" the results turn out as they do, many researchers and practitioners will "blame the trial," even multiple narrowly focused ones, and disregard the results.

In Project Ernestine, "CPM-GOMS models saved the field trial from a potential disaster" (Gray et al., 1993, p. 294) by providing the compelling explanation for the results. However, there are no modeling techniques in existence today powerful enough to explain the process by which a person or group using a UEM identifies usability problems. Thus, we need to look elsewhere for methods to explain the outcomes of UEM use.

Fallacy: Multiple, Narrowly Focused Controlled Experiments Are the *Only* Way to Gain Insight Into Advantages and Disadvantages of UEMs

The development of UEMs is still relatively new. The techniques examined in the studies that Gray and Salzman critique were developed within the last decade, with their major "how-to" publications and tutorials appearing only within the last 5 years. As Gray and Salzman say, "The development and definition of UEMs has been a dynamic enterprise. In fact, all currently used analytic UEMs have evolved rapidly over recent years" (p. 213). Thus, UEMs are in a formative stage of their development. As such, we need to consider using evaluation methods suitable for iterative design of these techniques, rather than summative comparison of them.

We have all been in this situation before. The system designer says: "It's too early to go for statistical significance—the design is inevitably going to change. Give us information to help change the design to make it better." So we do. HCI textbook after textbook (e.g., Hix & Hartson, 1993; Newman & Lamming, 1995; Nielsen, 1993; Preece et al., 1994; Shneiderman, 1998) advocate evaluation techniques for the formative stage of design that are relatively quick to do and that provide rich data to help designers understand problems and generate solutions to those problems. Because our UEMs are still in a formative stage, we should look for evaluation techniques that will give us the rich data necessary to track down the causes of problems with the UEM and help us improve them.

Perhaps the biggest difficulty I have with the studies critiqued by Gray and Salzman is that they provided no data about what the analysts actually did when they were using the UEMs; that is, they did not provide the necessary rich process data. Without process data, it is difficult to understand how the UEM itself leads the analyst to identify usability problems, as opposed to the analyst simply being clever. Without process data, a developer does not know what to expect when setting out to use a new UEM. Finally, without process data, it is difficult to provide meaningful feedback to the UEM developers so they can improve their techniques. The studies critiqued by Gray and Salzman are akin to a usability study of a system that only provides numerical data about performance time or total number of errors—a method the HCI community would never advocate to system designers in the early stages of development.

When a computer system is being designed, many textbooks recommends think-aloud usability tests (or some close variation) for collecting process data and understanding what users are doing and thinking. Nielsen (1993) said "thinking aloud may be the single most valuable usability engineering method" (p. 195). In addition to providing suitable data for formative design, few users are needed to be useful. Even Gray and Salzman remind us that "*experimental* results have shown that for *user testing*, only a few participants are needed to identify problems and even fewer participants are needed to identify severe problems (Virzi, 1992; although see J. R. Lewis, 1994, for a counter argument)" (p. 211). However, we cannot simply import think-aloud usability testing procedures to study the use of UEMs. UEMs typically take much more time to learn and use than would be tractable for collecting or analyzing think-aloud data.

In his book *Case Study Research: Design and Methods,* Robert Yin (1994) stated that a case study approach has an advantage over surveys, experiments, and other research strategies "when a 'how' or 'why' question is being asked about a contemporary set of events over which the investigator has little or no control" (Yin, 1994, p. 9). This seems to describe the situation currently facing the field of HCI when evaluating UEMs. We are asking how a given technique can be used to predict usability problems and why it works in some situations and not in others, and we have virtually no control over how an analyst learns or uses a technique. Thus, I suggest that we consider using case studies to understand the advantages and pitfalls of UEMs.

The essence of the case study research approach[1] is to collect many different types of data and use them "in a triangulating fashion" (Yin, 1994, p. 13) to converge on an explanation of what happened. When multiple sources of information converge, it boosts our confidence that we have understood the series of decisions occurring in a case: how these decisions were made, how they were implemented, and what result was achieved. This deep understanding should allow us to know whether these processes and results are likely to reoccur with other developers or in the next design project.[2] In addition, these data should enable us to identify problems with the UEMs themselves and suggest plausible solutions to those problems.

Lessons Learned in Preliminary HCI Case Studies

In reaction to several of the studies Gray and Salzman critiqued, I conducted some studies to gather richer data about how UEMs were used. I have learned several lessons from these first few studies, which I hope will help researchers interested in taking a similar path.

Content Analysis Aids Understanding. The first study was not a case study, but it concerns the types of data that would be useful in a case study. In it (John, 1994), I analyzed the content of Keystroke Level Models pro-

---

1. Do not confuse case study research with case study teaching. Cases for teaching are chosen to make pedagogical points, are often altered to make those points more clearly (Yin, 1994), and are often presented in such an informal way as to give the appearance of ad hoc or sloppy data collection.

2. The issue of generalization in a case study is not a simple one—knowledge of experimental techniques that depend on sampling a population to generalize to that population does not apply to case studies. In fact, Yin (1994) argued strongly that "a fatal flaw in doing case studies is to conceive of statistical generalization as the method of generalizing the result of the case ... the method of generalization is 'analytic generalization' in which a previously developed theory is used as a template with which to compare empirical results of the case study" (p. 31).

duced by eight students in replicating the results of Nielsen and Phillips's (1993) study. Although the numerical predictions of these students were similar to the numerical predictions of similar students in the Nielsen and Phillips paper, and both were similar to actual user performance, my analysis showed that the students frequently made two systematic errors in their models, and the errors happened to cancel each other out. In particular, the students tended to include more mental operators (Ms) than were recommended by Card, Moran, and Newell's (1983) procedure for placing Ms (as applied by their professor, me). Counteracting that, the students' models tended to be incomplete, leaving out whole steps. The additional Ms added time to the total estimate, but leaving out steps subtracted time, thereby, coincidentally, giving a near-to-correct estimate. This study demonstrated the added value of examining the content of analyses as well as the numerical results.

Diary-Based Studies Add Rich Process Data.   However, a content analysis does not give any insight into how that content was generated. What did these students know, think, and pay attention to that produced these errors? To alleviate that problem, I incorporated diaries into the studies of UEM use (John & Mashyna, 1997; John & Packer, 1995), modeled after the diary work done by John Rieman (1993). These diaries required the participant to indicate every half hour what category of activity he or she was engaged in (e.g., reading a paper, doing an analysis, reading and analysis so intertwined as to be inseparable, etc.) and to write free-form comments about difficulties with the UEM and insights into using the UEM. In addition, the participants used the diaries to record when usability problems were found with the system being evaluated, as well as structured Problem Description Reports (PDRs, borrowed from Jeffries et al., 1991) to record the usability problems themselves. Finally, the participants wrote reports summarizing their findings about the system and about the UEM.

These four pots of data—diaries, PDRs, final reports, and materials associated with the analyses themselves (e.g., GOMS models or Cognitive Walkthrough steps with their success or failure stories)—provide the rich data from which converging evidence of process can arise in a case study. For instance, strong links can be drawn from the activities recorded in the diaries, to the materials of the analyses, to the PDRs and final report, and back to the free-form comments in the diaries. Or, hypotheses can arise from a comment in the diaries and checked with reference to the PDRs (e.g., an early entry in the diary says "I'm concerned that I won't be able to keep track of all the usability problems I'll find" and indeed, many duplicate PDRs were submitted by the end of the analysis). This is not to say that a diary-based procedure is the only way to collect rich process data for a case study; other methods can clearly be useful (e.g., interviews, ethnography). However, some sort of rich process data greatly helps interpret outcome data.

Case Studies Generate Many Hypotheses. I have found that a case study generates many hypotheses that can be tested through further case studies or the narrowly focused controlled experiments Gray and Salzman favor. For example, our first case study of Cognitive Walkthrough (John & Mashyna, 1997; John & Packer, 1995) generated the hypothesis that Cognitive Walkthrough can be learned better from a single paper written specifically for practitioners (Wharton, Rieman, Lewis, & Polson, 1994) than from the suite of papers written for HCI researchers. Our second case study (Jacobsen & John, 1998) investigates that hypothesis with another participant with similar educational background, investing similar amounts of time in the analysis, but reading only that one how-to paper.

Case Studies Are Not Just Small Experiments! Unfortunately, many people have equated a case study with an experiment with too few participants to attain statistical significance. In fact, Gray and Salzman make this error in their comment about Jeffries's work:

> If Jeffries et al. (1991) had been cast as a case study (and appropriate changes made throughout), the paper would have provided a snapshot of the trade-offs facing Hewlett Packard in deciding how to do usability analyses in the late 1980s. Unfortunately, the work was presented as an experimental comparison of four UEMs, and several misleading conclusions were drawn. (p. 224)

A case study is not simply a small experiment cast in a different light. It requires several different types of data to allow triangulation, a plan for linking those data, theories to test, and many other details different from an experiment. This is recognized by prominent statisticians and experimentalists (in fact, precisely the experimentalists Gray and Salzman draw so heavily from), as evidenced by the following words of wisdom. "Certainly the case study as normally practiced should not be demeaned by identification with the one-group post-test-only design" (Cook & Campbell, 1979, p. 96; as cited in Yin, 1994, p. 19) and

> most people feel they can prepare a case study, and nearly all of us believe we can understand one. Because neither view is well founded, the case study receives a good deal of approbation it does not deserve. (Hoaglin, Light, McPeek, Mosteller, & Stoto, 1982, p. 134). (five "prominent statisticians" as quoted by Yin, 1994, p. 11)

Given these common misconceptions, a call for case studies in HCI research must be accompanied by a caution to learn to do them well. Gray and Salzman make the same point for experiments—the flaws they found in the studies they reviewed could have been "handled by following well-known experimental design considerations and reporting conventions" (p. 247). Likewise, let us avoid similar problems with case studies by adhering to already established procedures for case study design, execution, and interpretation. Yin's 1994 book is an excellent place to start.

Conclusion

In summary, this commentary calls for a consideration of a broader range of evaluation methods for UEM than the "narrowly focused experiments" advocated by Gray and Salzman. In particular, I suggest the use of case studies that collect content and process data as well as outcome data. However, just as Gray and Salzman's article presents a cautionary tale about conducting valid experimental research, caution should be used when conducting or reading (or reviewing!) case studies as well. Valid conclusions require education in the method and careful application of that knowledge to the evaluation at hand.

References

Card, S. K., Moran, T. P., & Newell, A. (1983). *The psychology of human–computer interaction.* Hillsdale, NJ: Lawrence Erlbaum Associates, Inc.

Cook, T. D., & Campbell, D. T. (1979). *Quasi-experimentation: Design and analysis issues for field settings.* Chicago: Rand McNally.

Gray, W. D., John, B. E., & Atwood, M. E. (1993). Project Ernestine: Validating a GOMS analysis for predicting and explaining real-world task performance. *Human–Computer Interaction, 8,* 237–309.

Hix, D., & Hartson, H. R. (1993). *Developing user interfaces: Ensuring usability through product and process.* New York: Wiley.

Hoaglin, D. C., Light, R. J., McPeek, B., Mosteller, F., & Soto, M. A. (1982). *Data for decisions: Information strategies for policymakers.* Cambridge, MA: Abt.

Jacobsen, N., & John, B. E. (1998). *A tale of two critics: Case studies in using cognitive walkthrough.* Manuscript in preparation.

Jeffries, R., Miller, J. R., Wharton, C., & Uyeda, K. M. (1991). User interface evaluation in the real world: A comparison of four techniques. *Proceedings of the ACM CHI'91 Conference on Human Factors in Computing Systems,* 119–124. New York: ACM.

John, B. E. (1994). Toward a deeper comparison of methods: A reaction to Nielsen & Phillips and new data. *Proceedings Companion of the ACM CHI'94 Conference on Human Factors in Computing Systems,* 285–286. New York: ACM.

John, B. E., & Mashyna, M. M. (1997). Evaluating a multimedia authoring tool with cognitive walkthrough and think-aloud user studies. *Journal of the American Society of Information Science, 48,* 1004–1022.

John, B. E., & Packer, H. (1995). Learning and using the cognitive walkthrough method: A case study approach. *Proceedings of the ACM CHI'95 Conference on Human Factors in Computing Systems,* 429–436. New York: ACM.

Lewis, J. R. (1994). Sample sizes for usability studies: Additional considerations. *Human Factors, 36,* 368–378.

Newman, W. M., & Lamming, M. G. (1995). *Interactive system design.* Wokingham, England: Addison-Wesley.

Nielsen, J. (1993). *Usability engineering.* Boston: Academic.

Nielsen, J., & Phillips, V. L. (1993). Estimating the relative usability of two interfaces: Heuristic, formal, and empirical methods compared. *Proceedings of the ACM/IFIP CHI'93/INTERACT'93 Conference on Human Factors in Computing Systems,* 214–221. New York: ACM.

Preece, J., Rogers, Y., Sharp, H., Benyon, D., Holland, S., & Carey, T. (1994). *Human–computer interaction.* Wokingham, England: Addison-Wesley.

Rieman, J. (1993). The diary study: A workplace-oriented research tool to guide laboratory efforts. *Proceedings of the ACM/IFIP CHI'93/INTERACT'93 Conference on Human Factors in Computing Systems,* 321–326. New York: ACM.

Shneiderman, B. (1998). *Designing the user interface: Strategies for effective human–computer interaction.* Reading, MA: Addison-Wesley.

Virzi, R. A. (1992). Refining the test phase of usability evaluation: How many subjects is enough? *Human Factors, 34,* 457–468.

Wharton, C., Rieman, J., Lewis, C., & Polson, P. (1994). The cognitive walkthrough method: A practitioner's guide. In J. Nielsen & R. L. Mack (Eds.), *Usability inspection methods* (pp. 105–140). New York: Wiley.

Yin, R. K. (1994). *Case study research: Design and methods* (2nd ed., Vol. 5: Applied Social Research Methods Series). Thousand Oaks, CA: Sage.

---

6. Experiments Are For Small Questions, Not Large Ones Like "What Usability Evaluation Method Should I Use?"

Andrew F. Monk

The author is a Reader in Psychology at the University of York, UK; he is currently researching electronically mediated communication, especially video as data. Address: Department of Psychology, University of York, York, YO1 5DD, UK. E-mail: am1@york.ac.uk.

Gray and Salzman are quite right to criticize the conclusions drawn in the four studies they discuss. The problems they identify are not limited to studies of usability evaluation methods (UEMs). The misuse of experiments is widespread throughout HCI. It arises from a naive belief among computer scientists and psychologists that numbers somehow reflect "the truth." There is a value for carefully constructed experiments reporting quantitative results, but it is not to answer grand questions such as those considered by Gray and Salzman. When they state that "The good news is that none of the problems we found are unique to HCI and all can be overcome, or at least mitigated ... " (p. 247) they are deluding themselves. This commentary is to argue that it will never be practical to compare UEMs in the way that engineers or managers want and, more generally, that HCI researchers need to change the way they think about the purpose of experimentation.

When I was young I ran experiments on visual word recognition. These are the kind of experiments Gray and Salzman are trying to emulate. Each experiment asked one or two very small and specific questions; most of these were very small and specific questions about another experiment. I was interested in the word superiority effect, a collection of findings suggesting that it may be easier to perceive words than the individual letters contained in them. Baron (1978) reviewed 58 papers on this topic published since 1954. The way such a literature evolves is as follows: X publishes an experiment that seems to demonstrate a relevant phenomenon and suggests an interpretation of why it should have occurred. Y thinks of an alternative explanation and publishes an experiment testing that explanation. Z then tests an alternative interpretation of Y's result, and so on. Ideally, the experiment constructed by Y differs only minimally from that constructed by X—for example, changing the words used, the way that performance is measured, or even something as apparently subtle as the instructions given to the participants. In this way, over a period of over 20 years, a gradual understanding of the word superiority effect developed, leading in the late 1970s to some sophisticated and subsequently influential computer models of visual word recognition (McClelland & Rumelhart, 1981).

The reviewers and editors employed by the journals publishing these studies ensured that all the methodological points discussed in Gray and Salzman's article were covered. Participants for the experiment are drawn from the limitless pool of students at British and American universities, and so there was no problem ensuring sufficiently large sample sizes for the statistical techniques employed. Statistical tests are necessary because of the large amount of error variance in behavioral studies. People are different and will perform differentially even though they are in the same treatment group. In addition, there is always great deal of measurement error attendant with behavioral measures. These two sources of noise in the data means that there is always the possibility that a result has occurred by chance; for example, random error variance can cause one group to have higher scores in it than another group. Statistical tests work by assessing the possibility that a result could have occurred by chance. By making certain assumptions, the probability that this is the case is computed. If that probability is very small, the possibility is rejected and the result is said to be significant. In studies of word recognition, statistical conclusion validity is examined taking account of variation across participants and across materials. For example, in one of my experiments (Monk & Hulme, 1983, Experiment 1) there were 60 participants and 40 words. The characteristics of both the words ("materials") and participants were carefully specified. Just as we needed to demonstrate that chance and the inevitable variations in performance between participants could not account for the results, we also needed to demonstrate that chance and the inevitable variation within our sample of words could not account for the results. Sophisticated statistical techniques are required to do this (Clark, 1973).

The other problems of internal, external, and construct validity described by Gray and Salzman are made tractable by asking only very small questions and drawing only very small conclusions. As was explained previously, most of the questions asked in these experiments on visual word recognition related directly to someone else's experiment; thus one can be very confident about what has been manipulated. If the small change made by Y to the method used by X gives a different result, one can be fairly sure that the result was caused by that change. Framing questions and conclusions with regard to a previous experiment also serves to strictly limit the scope of the generalization required. One only needs to be able to say that the result should be repeatable in another similar laboratory context. There is no requirement to generalize to the "real" world. The interpretations drawn from the findings may have wider implications, but these implications are only rehearsed when attempting to get funds for one's research and not when describing its practice in a journal article!

So why can't this model be applied to comparisons between UEMs or in HCI research more generally? The answer has to do with the generality of the findings required and the practical constraints on achieving that generality. The logic behind significance testing was explained previously. One further principle needs to be explained to understand how statistical tests may help in the question of generalization. When carrying out market research, for example, the first step is to specify the target population one wishes to generalize to. This might be "British adult males between age 20 and 40." One then goes to some trouble to sample randomly from this population. Statistical reasoning can then be used to calculate how confident one may be generalizing conclusions drawn about the sample to the target population as a whole. In the word recognition experiments described previously, students were not sampled at random from the population of all students. They were simply recruited and assigned at random to the experimental groups. The statistics used indicate that it is very unlikely that our conclusions could have arisen by chance from this random allocation of students to groups. They do not allow us to generalize from this sample of students to other students or humanity in general. To do this, one must appeal to the usually implicit assumption that "for the effects being considered in this experiment" one human being is much like another. York students who volunteer for experiments are simply assumed to have similar word recognition processes to all other adult readers of English. This assumption is easily tested and is reasonable to accept until it is proved to be false.

Of course, the market researcher cannot make the same assumptions about, say, attitudes to consumer products. Neither can an HCI researcher assume that effects observed when students use two UEMs will be the same as when software engineers use them. Computer science students are sometimes used as substitutes for software engineers (e.g., Wright & Monk, 1991), but most studies attempt to recruit participants from the ranks of industry. Even then it is not clear how homogenous the target population is. Most would expect a distinction between software engineers and human factors specialists, but are the software engineers that you happen to have access to typical of all software engineers? Perhaps the company recruited them in an unusual way, or perhaps the projects they have worked on as a group makes them different from software engineers in other divisions of the company. As the experimental tasks used get closer to real work, these concerns become more salient. One can't get much closer to real work than the task of using a UEM, and so one would expect that the participant's training and experience to strongly influence any effects observed.

The real scope of a typical UEM study is thus very narrow. Gray and Salzman give as an example of tractable question

Does UEM-A find more feedback problems in walk-up and use interfaces than UEM-B?

Really the question is

Does UEM-A find more feedback problems in walk-up and use interfaces than UEM-B when they are used by the engineers employed in 1997 in Division X of Company Y?

To generalize more broadly with confidence, the market research strategy must be employed. Rather than recruiting participants through serendipitous contacts, mutual favors, and so on, a target population should be defined, such as "developers working on database products in companies in the United States." Individuals should then be sampled at random from this population. One must further assume that any bias introduced by those sampled who refuse to take part is not important. To demonstrate this, one should collect demographic data about the participants used, their years of experience, age, qualifications, and so forth, and show that the average and range are similar to those for the target population to which one is generalizing. The practical and "political" issues discussed later in this commentary mean that this is seldom, if ever, going to happen.

This is only the beginning of the generalization nightmare. In the word recognition experiments some trouble was taken to specify clearly the materials used. Words were selected with well-specified criteria and statistical tests applied to evaluate the possibility of chance effects arising from this source. The full list of words used is given in an appendix so that readers can apply their own measures to characterize the selection made. Gray and Salzman suggest that it is best to apply a UEM to more than one software package, but is this enough? If just two UEMs are used the scope of the study becomes:

> Does UEM-A find more feedback problems in walk-up and use Interfaces P and Q than UEM-B when they are used by the engineers employed in 1997 in Division X of Company Y?

Alternatively, the population of walk-up and use interfaces to be generalized to must be characterized in some way and an argument made that P and Q are typical of that class of interfaces. Better, define the target population of software packages, sample at random from it, and then perform the relevant statistical tests.

A related practical problem is training. Realistic work tasks, such as using a UEM, require realistic amounts of training. In a word recognition experiment it is practical to train participants as a part of the experiment. In contrast, the participants in an experiment comparing two UEMs will have to be trained in a way that fits in with the way that they work. This results in some of the differences in setting and the different ways that a single UEM may be realized in different settings, as discussed by Gray and Salzman. Again, either one reduces the scope of the question asked still further or one defines the target population of settings one wishes to generalize to and samples at random from it. The latter will seldom, if ever, be practical.

In summary, the real scope of a question that may be practically asked in a study of two UEMs is this:

> Does UEM-A find more feedback problems in walk-up and use Interfaces P and Q than UEM-B, when used with the following modifications, in the setting specified, by the engineers employed in 1994 in Division X of Company Y, trained as specified?

Further generalization is impractical because of the way these studies come about. Getting access to busy practicing software engineers or usability professionals is extremely difficult. Someone quite powerful in the organization will have to sanction the considerable expense to the organization of taking these people away from their daily work. The real triumph of the four studies criticized by Gray and Salzman is that they achieved this at all. Random assignment of methods and software packages to participants, strict control over training, setting, and experience all add to the expense to the organization. The payoff for the organization cooperating in such a study is limited anyway. The known effectiveness of a UEM is only one among many criteria by which a company decides whether to use the UEM. Issues concerning staff selection and training, or how easy it is to fit it into existing work practices, will generally take preference. Further, there is no payoff for the organization in knowing that the findings generalize to other organizations. Indeed, they may be openly hostile to the idea that other organizations will get early information that the study is taking place.

The cautions described here and by Gray and Salzman apply to most experiments in HCI. The arguments are not new and were rehearsed in the discussion of the use of quantitative user testing methods for the evaluation of prototypes (Carroll, 1990; Monk & Wright, 1991). Neither will general conclusions develop incrementally over the course of numerous studies as they did in the word recognition example. There will never be 58 studies of a single phenomenon in HCI because the areas of interest change as technologies evolve, and there are relatively few investigators for the size of the discipline. This is not to say that experimentation in HCI is without value. There is value in demonstrating that, at least under some circumstances, some proposition is true. The exercise of demonstrating that proposition in an experiment, and being honest about what the experiment actually tells you, is very informative. The real value of an experiment is that it forces the investigator performing the experiment, and the reader of a paper describing it, to think clearly about important issues. Gray and Salzman's discussion, for example, raises several fundamental issues that have to be addressed. To take just three:

- What is usability? The act of operationalizing aspects of usability as quantitative measures and justifying their use against the criteria specified by Gray and Salzman forced a clarity of thought that had previously been absent. Developing new measures and demonstrating their use in experiments will further clarify our thinking.
- What defines a particular UEM? Similar advantages accrue to characterizing the different ways a UEM may be used and how this variation is to be controlled in an experiment. This will require field studies of real work. A clear understanding of how a UEM is really used in prac-

tice is of considerable general value. It may allow one to provide better training, as an example.

- What are the use characteristics of different software packages? The problem of generalizing from the software package tested to other packages forces the investigator to think about ways of classifying and characterizing them. An effective taxonomy that satisfies the needs of experimentation would also have general value. It could be one input when selecting UEMs for real evaluations, for example.

What is being suggested here is a different view of the purpose of experiments in HCI from that implied by Gray and Salzman. Experiments should be employed to answer small questions, and HCI researchers should not be embarrassed about the limited scope of the conclusions they draw. Defining that scope will elicit important issues with wide conceptual import. A potential danger is that the wrong narrow questions are asked so that no practically important issues are raised. A potential strategy for avoiding this problem is to use experimentation in combination with field studies. At York (for an example see Daly-Jones, Monk, Frohlich, Geelhoed, & Loughran, 1997) we are following a research strategy that starts with the identification of a particular technological innovation and a potential setting for its application. Field studies of real work are then used to identify broad issues of relevance to the implementation of the technology in that setting. One or two of these issues are then selected for intensive investigation in experiments. The experiments serve to clarify the conceptual basis of the issue. These insights can then be taken back into the real world as concepts to aid effective design.

References

Baron, J. (1978). The word superiority effect: Perceptual learning from reading. In W. K. Estes (Ed.), *Handbook of learning and cognitive processes: Vol 6. Linguistic functions in cognitive theory* (pp. 131–166). Hillsdale, NJ: Lawrence Erlbaum Associates, Inc.

Carroll, J. M. (1990). Infinite detail and emulation in an ontologically minimized HCI. *Proceedings of ACM CHI'90 Conference on Human Factors in Computing Systems*, 321–327. Reading, MA: Addison-Wesley.

Clark, H. H. (1973). The language as a fixed-effect fallacy: A critique of language statistics in psychological research. *Journal of Verbal Learning and Verbal Behaviour, 12*, 335–359.

Daly-Jones, O., Monk, A. F., Frohlich, D., Geelhoed, E., & Loughran, S. (1997). Multimodal messages: The pen and voice opportunity. *Interacting With Computers, 9*, 1–25.

McClelland, J. L., & Rumelhart, D. (1981). An interaction activation model of context effects in letter perception: Part 1. An account of basic findings. *Psychological Review, 88*, 357–407.

Monk, A. F., & Hulme, C. (1983). Errors in proofreading: Evidence for the use of word shape in word recognition. *Memory and Cognition, 11*, 16–23.

Monk, A. F., & Wright, P. C. (1991). Observations and inventions: New approaches to the study of human–computer interaction. *Interacting With Computers, 3*, 204–216.

Wright, P. C., & Monk, A. F. (1991). A cost-effective evaluation method for use by designers. *International Journal of Man–Machine Studies, 35*, 891–912.

---

7. What's Science Got to Do With It? Designing HCI Studies That Ask Big Questions and Get Results That Matter

Sharon L. Oviatt

The author is Associate Professor of Computer Science in the Center for Human Computer Communication at the Oregon Graduate Institute of Science and Technology; her current interests include multimodal HCI and interface design for spoken language systems, multimodal systems, and portable technology.

In their critique of five experimental studies comparing different usability evaluation methods (UEMs), Gray and Salzman provide a stimulus for HCI researchers to reflect on methodology from both a technical and stylistic point of view. Because any analysis of methods only makes sense in light of specific goals, their critique also prompts consideration of those research goals that the HCI community deems worthy of investigation. Although the methodological concerns raised in this critique focus on specific studies comparing UEMs, they are not specific to this area of research. The comments that follow offer further perspective on a few of the more substantial issues raised by Gray and Salzman's critique and the studies they review.

Asking Questions and Establishing Trade-Offs

Perhaps the most important point made by Gray and Salzman is that studies comparing UEMs "must be conducted that delineate the trade-offs—the advantages and disadvantages—of each method" (p. 207). That is, it is unsatisfactory to ask in a generic or unqualified way "whether UEM-A is better than UEM-B" (p. 244). The basic point is that different UEMs are to a large extent apples and oranges—they are capable of telling us different things about usability. A more interesting and informative collection of scientific questions instead might be the following:

- What kinds of problems typically are uncovered by different UEMs—including specific and general problems, expert-defined and user-defined problems?
- Under what circumstances do these different types of problems tend to be uncovered by a given UEM—for what kinds of systems, users, and usage contexts?
- At what point in the design cycle are different UEMs most valuable in uncovering important problems?
- What adverse consequences would these problems have had for usability, both from the designers' and users' point of view?
- How can different UEMs be used *in combination* with one another to identify problems that threaten the usability of a given system?

The last question is particularly important, because individual methods seldom are used as stand-alone techniques when answering real-world questions in corporate labs or in the field. In this respect, *methodological diversity* is our richest resource and the use of *multiple converging methods* our most powerful design technique.

The Importance of Converging Methods and Metrics

Given this perspective, the UEM research agenda might become more meaningful and informative by asking further questions, such as the following:

- Which UEM methods ideally should be combined during evaluations, and why is their combination optimal?
- What cumulative usability problems can these UEM combinations uncover?
- What interdependencies exist between the identification of different usability problems—does identifying one problem with UEM-A increase the likelihood of identifying a second problem with UEM-B?
- When two different methods both identify the same problem during testing, does this tend to produce a deeper or different understanding of the nature of the usability problem?
- Does this altered view of the problem then also lead to a different usability solution?

Regarding dependent measures, Gray and Salzman critique the studies in question for tallying "usability problem counts" as their standard metric for comparing UEMs, in part with the goal "of providing focused feedback to software designers on specific problems that if fixed would increase usability" (p. 247). However, they assert that such counts have been relied on too heavily as a singular metric for comparing UEMs and then for prescribing one UEM as better or more efficient than another. Gray and Salzman question whether "the UEM that names the most potential problems is the most effective" (p. 247). Although a useful tool, they argue that such counts are by nature limited—just as any single research method is limited. Furthermore, problem counts alone would not provide sufficiently rich data to establish fuller trade-off information regarding the strengths and weaknesses of different UEMs.

There are several possible ways of expanding on this limited dependent measure to provide more meaningful information and to circumvent these criticisms. For example, usability problems could be assessed more comprehensively by recording the problems that are perceived and reported by *users*, as well as those identified by expert testers. If user testing was conducted to collect such data, then performance measures also could be summarized. Likewise, subjective self-report data could be collected on users' and expert testers' view of the nature and severity of problems encountered and the value of different UEMs.

Abandoning Self-Defeating Myths That HCI Cannot Be Science

One option for understanding the uses of different UEMs might be to design a richer and more eclectic study in which individual and combined UEMs are compared. Such a study could include a diverse range of both qualitative and quantitative measures. Furthermore, it would be valuable to design a longitudinal study in which motivated novice professionals are trained in the workplace on the UEMs of interest and then testing is conducted to evaluate the utility of alternative UEM techniques when used individually (UEM-A, UEM-B, etc.) versus in potentially effective *combinations* (UEMs-AC). Such a study also would facilitate exploring the ease with which different UEM techniques are learned and could help to elucidate the process of applying UEMs effectively in the workplace. The point is that a more in-depth study could generate valuable insights into the strengths, weaknesses, and best uses of alternative UEMs, which could help to establish a richer understanding of UEM methods and the tradeoffs entailed in using them successfully.

Needless to say, any rich longitudinal study of this kind would be labor intensive and a considerable investment, so accomplishing it might be effectively done by teaming between industrial and academic researchers. Such research certainly could be accomplished in a manner designed to establish cause-and-effect relations with reasonable precision, while focusing on a meaningful and broad range of HCI issues. Such an effort need not be defined as a "parochial" experimental enterprise, and the inclusion of powerful experimental techniques does not necessarily mean that the research can only be focused on "small questions" or "irrelevant issues." In short, it is time to abandon any self-defeating myths implying that HCI research is not the domain of "real science." Although posing a methodological challenge, it is our job as HCI researchers to be creative about developing appropriate empirical methods for successfully asking and answering meaningful scientific questions about important HCI topics.

The Importance of Replication

Another important methodological theme in Gray and Salzman's critique involves establishing the generality of experimental work comparing UEMs. Gray and Saltman emphasize the elementary but important point that "a well-conducted, valid experiment permits us to make strong inferences regarding two important issues: (a) cause and effect and (b) generality" (p. 208). This introduction to the inferential power of experimentation then is followed with an extensive critical focus on the design shortcomings that prevented the present UEM studies from demonstrating experimental generality. Although they outline these shortcomings, Gray and Salzman acknowledge that "few studies are generalizable across all times and places" (p. 208). Perhaps a more accurate statement would have been that it is *impossible* for any single experimental study to demonstrate the *complete generality* of its findings. Although many of the methodological points outlined in Gray and Salzman's review are quite well taken, their emphasis on self-contained experimental demonstrations of generality is an unreasonable expectation.

The usual scientific expectation is that any consequential research findings will be recognized as such by the broader HCI community, who will make it a priority to replicate them—in fact sometimes repeatedly, as in the case of prediction of movement time by Fitts' Law in different HCI contexts (Card, English, & Burr, 1978; MacKenzie & Buxton, 1992; Ware & Mikaelian, 1987) or prediction of disfluent and hyperarticulate speech in spoken language interfaces (Oviatt, 1995, 1997; Oviatt, MacEachern, & Levow, 1998). Replication of experimental work is a particularly powerful tool because it permits research to establish the generality of findings across different researchers, environments, task domains, users, and systems. If two independent studies are able to replicate an important finding, then the likelihood that the results in question could have co-occurred by chance alone is reduced dramatically. This scientific replication process takes time, multiple experiments, and usually work by more than one researcher within the community. To establish the strongest case for generality, in fact, it is preferable for more than one researcher and laboratory to participate in an experiment's replication. One rather fortunate implication of this is that no researcher is single-handedly held accountable for completing the process of establishing experimental generality.

Conclusion

Gray and Salzman's critical review of studies comparing UEMs provides one stepping stone that suggests how this research topic could be pursued in the form of more sophisticated experimental studies. In addition, the preceding brief commentary also asks leading questions and offers ideas that inject a different perspective on structuring future research. However, the ultimate stimulus for improving research on UEMs would be for some innovative researcher to conduct a model study—one that builds upon the foundation of the early studies reviewed here, that demonstrates tangible experimental design improvements on them, and that generates richer and more provocative data on a topic that clearly is both important and controversial.

References

Card, S. K., English, W. K., & Burr, B. J. (1978). Evaluation of mouse, rate-controlled isometric joystick, step keys, and text keys for text selection on a CRT. *Ergonomics, 21,* 601–613.

MacKenzie, I. S., & Buxton, W. (1992). Extending Fitts' Law to two-dimensional tasks. *Proceedings of the CHI'92 Conference on Human Factors in Computing Systems,* 219–226. New York: ACM.

Oviatt, S. L. (1995). Predicting spoken disfluencies during human–computer interaction. *Computer Speech and Language, 9*(1), 19–35.

Oviatt, S. L. (1997). Multimodal interactive maps: Designing for human performance. *Human–Computer Interaction, 12,* 93–129.

Oviatt, S. L., MacEachern, M., & Levow, G. (1998). Predicting hyperarticulate speech during human–computer error resolution. *Speech Communication, 24*(2), 1–23.

Ware, C., & Mikaelian, H. H. (1987). An evaluation of an eye-tracker as a device for computer input. *Proceedings of the CHI + GI'87 Conference on Human Factors in Computing Systems and Graphics Interface,* 183–188. New York: ACM.

8. Review Validity, Causal Analysis, and Rare Evaluation Events

John M. Carroll

The author is a cognitive/computer scientist interested in learning and problem solving in HCI contexts and in methods and tools for instruction and design; he is currently Professor of Computer Science and Psychology, Director of the Center for Human–Computer Interaction, and Head of the Computer Science Department at Virginia Tech. Address: 660 McBryde Hall, Virginia Tech, Blacksburg, VA 24061–0106. E-mail: carroll@cs.vt.edu.

The investigation of usability evaluation methods (UEMs) is a topic of great importance to HCI. Gray and Salzman provide solid methodological guidance for improving experimental studies of UEMs. Nevertheless, their article is somewhat parochial and severe both in content and tone. To adopt Gray and Salzman's quirky language, I think these distortions constitute "threats" both to the constructive impact of their article and to the development of an effective and well-grounded engineering practice in HCI.

Gray and Salzman go far beyond what their data warrant in indicting the entire field of HCI for ignorance of and disinterest in the design of experiments. In doing this, they ironically demonstrate the methodological practices for which they criticize others. They reviewed only five somewhat dated conference papers. The role of conference papers is rapid dissemination of new ideas, experiences, and results. The five papers played this role, triggering a lively technical debate in which Gray and Salzman are only the most recent of many participants. Conference papers should take risks, be controversial, be tentative, and be incomplete. It is certainly fair to critically revisit technical work, but it is not legitimate to assert or to imply that five elderly conference papers represent the methodological standard of the entire field.

Leaving aside questions about the validity of their analysis, Gray and Salzman present a somewhat austere view of methodology and inference. Simply put, they regard experiments as the royal road to causal analysis, biblically quoting and requoting Cook and Campbell (1979): "the unique purpose of experiments is to provide stronger tests of *causal* hypotheses than is permitted by other forms of research" (p. 83). I think that this is too narrow a program for HCI. For example, simulation studies provide more powerful, articulate, and cost-effective tests of causal hypotheses than do experiments. Reading beyond Gray and Salzman's text, it seems that some of the difficulties in the UEM area stem from overly crude taxonomizing of the methods themselves. The methods should perhaps themselves be analyzed instead of being regarded merely as levels of a single independent variable. This is a traditional shortcoming in experimental human factors research and the kind of limitation simulation studies could address.

Another example is the trade-off between rigor and relevance. It is not uncommon to value rigor over relevance, but we should not do so merely out of compulsion. Indeed, Gray and Salzman lament that the usability methods papers they regard as experimentally sound have had less influence than the five they critique. They reflect on this, suggesting that the broad scope of the five papers they reviewed was a key to their greater impact. However, perhaps broad scope is a critical attribute of influential HCI method studies. If so, do we really wish to encourage only rigorous studies that fail to have this attribute?

We should recast Gray and Salzman's enterprise more constructively as raising the question of how to appropriately manage rigor and relevance in usability methods research. There is methodology literature beyond Cook and Campbell that may provide guidance to us in doing this (e.g., Argyris, 1980; Schön, 1983). HCI should not court the fate of postwar operations research: Faced with the failure of its models in crucial but broadly complex areas like business management, housing policy, and criminal justice, the field shifted attention to simpler problems (Schön, 1983, pp. 16, 42–44). Although we must be aware of methodological issues and trade-offs, we cannot adjust the requirements of the world to our models; we must in fact do the reverse.

In their focus on conventional experiments, Gray and Salzman underestimate the importance and the distinctiveness of rare evaluation events. They criticize UEMs for "naming" more usability issues than an empirical survey identified. Statistically rare but critical events are part of the motivation for empirical approaches like the critical incident method (Flanagan, 1956) and ethnomethodology (Suchman, 1987). Indeed, the high-frequency events Gray and Salzman might prize as well validated (e.g., overpracticed keystroke-level behaviors) are often among the least important usability phenomena. Labeling rare but potentially critical data as false alarms is a misleading metaphor.

I was disappointed with the cynical tone that Gray and Salzman sometimes adopt. I do not like the implication that researchers are hucksters. At one point Gray and Salzman assert that a researcher "capitalized" on low statistical power (why not say he exposed his conclusions to additional risk?). Their title, *Damaged Merchandise,* makes it seem that the field is populated by salesmen and consumers and that the latter should get their money back! I think this kind of cynicism undermines the camaraderie among researchers and practitioners that is essential to the success of HCI.

It is important to have an ongoing discussion about methodology. I hope the Gray and Salzman article will encourage broader, bigger, more rigorous

UEM experiments. However, I also hope it will not discourage other complementary approaches to developing and assessing UEMs. We need them all.

References

Argyris, C. (1980). *Inner contradictions of rigorous research*. New York: Academic.
Cook, T. D., & Campbell, D. T. (1979). *Quasi-experimentation: Design and analysis issues for field settings*. Chicago: Rand McNally.
Flanagan, J. C. (1954). The critical incident technique. *Psychological Bulletin, 51*(28), 28–35.
Schön, D. A. (1983). *The reflective practitioner: How professionals think in action*. New York: Basic Books.
Suchman, L. A. (1987). *Plans and situated actions*. New York: Cambridge University Press.

---

9. Triangulation Within and Across HCI Disciplines

Wendy E. Mackay

The author is a visiting professor at the University of Aarhus. She is currently working in augmented reality and user innovation. Address: Department of Computer Science, University of Aarhus, Aabogade 34, DK–8200 Aarhus N, Denmark. E-mail: mackay@daimi.au.dk.

How should we evaluate interactive software? Gray and Salzman examine five influential studies that compare methods of evaluating usability. In a carefully argued critique, they identify numerous threats to validity and point out serious flaws in both the designs and claims made in each study. I think there will be various reactions to this article. Some people will react with alarm, concerned that if frequently cited articles in respected publications are so full of errors, perhaps the field of HCI is in serious trouble. Others may attack the article, misinterpreting it as the blind application of laboratory-style experiments to field studies. My own reaction is to commend Gray and Salzman for daring to critique some of the most influential work in the field, not only showing how to identify certain kinds of problems, but also showing how to address them. Their study helps to advance the field by raising the level of critical analysis in a graceful and constructive manner. This is how research disciplines make progress: through self-analysis and the creation of a solid research context.

Field Studies Versus Laboratory Experiments

Like many others in HCI, I started out with academic degrees in experimental psychology. Moving into industry, I quickly realized that laboratory experiments were rarely useful for addressing the HCI problems we were interested in. We borrowed and reinvented design and evaluation methods, keeping what worked and exploring ways of improving them. It was not until I later returned to academia for a PhD in a different field (management of technological innovation) that I discovered the wealth of relevant literature from other social sciences. My well-thumbed copy of Cook and Campbell (1979) proved far more useful than my aging psychology texts and research articles. As a psychologist, I was conditioned to believe that other social sciences were not particularly scientific. I discovered that there exists a solid foundation for conducting field studies, grounded in scientific principles but tempered with humility: The constraints of the real world make this type of research difficult to do well.

For those readers who are discouraged after reading Gray and Salzman's critique, I strongly recommend Cook and Campbell (1979) for concrete advice on how to conduct and analyze quasi-experiments or field studies. They begin by explaining the philosophical history of causal inference and the scientific method, including various measures of validity. The rest of the book is devoted to descriptions of quasi-experimental designs and modes of analyzing the data that result from them. For each design, they pointed out the most likely sources of invalid conclusions and offer suggestions for how best to avoid them. Threats to validity exist in any research setting. Cook and Campbell offer methods for conducting scientifically grounded research, given the variety of constraints found in real-world settings.

Building a Scientific Research Context for HCI

All natural sciences assume that individual studies build upon each other: No single study can ever address all possible threats to validity or potential biases by the researchers. Because individual studies are never viewed as conclusive, subsequent studies are required to present new results and to replicate or challenge earlier findings. Multiple studies, run by different people in different settings, help to reduce the number and severity of threats to validity and increase the strength and power of the results.

HCI, as a field, lacks an agreed-upon research context. Individual studies rarely replicate results from previous studies, much less build upon them. This is partly due to its youth: It takes time to develop a shared research foundation. This is also due to the multidisciplinary nature of the field: Researchers trained in one discipline (such as computer science) often feel compelled to perform studies outside of their expertise (such as laboratory experiments) for their work to be accepted. Even people with scientific training (such as experimental psychologists) may not be trained in research techniques that are most relevant to the problem at hand (i.e., field studies). Directly applying laboratory methods in field settings gives poor results, yet providing only anecdotal evidence misses an opportunity to discover general principles. HCI researchers must develop a set of research techniques that apply to the problems specific to HCI, either borrowed from other fields or invented for the purpose.

Another critical aspect of a shared research context is the peer review process. This is particularly complex in HCI, with people trained in many disciplines, sometimes with little or no overlap. For example, we do not have a good notion of "expertise." Is a computer scientist with extensive experience in graphical user interfaces but no background in statistics considered an expert reviewer of an experiment comparing different graphical interfaces? Should a university statistics course taken a decade ago be considered adequate background for critiquing an experimental paper? Asking people with diverse backgrounds to review a paper helps to ensure that the resulting papers are accessible to a wider, more diverse audience. It also avoids a common phenomenon in many sciences, in which tiny groups of researchers concentrate on increasingly detailed aspects of a problem and end up speaking only to themselves. On the other hand, we need a better classification of expertise—distinguishing between expertise in the discipline, in the specific content, and in the development or evaluation techniques used. We must agree on standards for particular types of papers and try to ensure that papers are reviewed by people with the necessary technical expertise. We must also learn to value papers with muted claims, rather than looking only for "exciting" results that may simply be the result of a less-than-careful analysis.

Triangulation Within Research Studies

For HCI to mature as a field, we must develop a shared research context, despite our internal diversity. Replication of studies is not enough. We need triangulation: using different techniques to operationalize behavior (i.e., specifying specific, measurable actions) while attempting to measure the same phenomenon. Gray and Salzman describe how Bailey, Allan, and Raiello (1992) used triangulation when they conducted two independent experiments to evaluate Molich and Neilsen's (1990) Heuristic Evaluation technique: "A strength of this report is that Experiment 2 essentially replicates the Experiment 1 findings using a different style of interface. This replication greatly increases the generality and construct validity of the findings" (p. 241).

Another type of triangulation involves examining different types of data gathered within the same study. For example, I conducted a controlled study of an "intelligent tutor" for a text editor in an industrial setting. In the early 1980s, applying artificial intelligence to online teaching was all the rage. We wondered just how much "intelligence" was necessary. Before actually implementing our tutor, we used a Wizard of Oz technique to decide.

We used a within-subjects design, providing identical amounts of tutoring to pairs of participants. For one third of the editing commands, the wizard would watch 1 participant performing a specified task and send a tutoring message when it was relevant to the task the participant was *currently* performing. The other participant of the pair would receive the same tutoring message, independently of what she was doing. For the second set of editing commands, the researcher would focus on the other participant in the pair, providing relevant tutoring messages to her and effectively random advice to the first participant. No tutoring was given for the remaining third of the editing commands. The quantitative results were interesting: Participants learned both sets of tutored commands, regardless of whether or not the tutoring was related to their current behavior. (In the control condition, participants learned very few of the untutored commands.) The purely quantitative analysis suggested that participants did not "need" an intelligent tutor: Random presentation of advice was equally effective. However the qualitative results were quite different. All but one of the participants reported that they really enjoyed receiving tutoring when it was contingent on their behavior and disliked it when it appeared randomly.

By triangulating, we gained a deeper understanding of what was going on. However, the study still leaves major design questions. Do we decide to ignore the qualitative data and simply implement a random tutor because it is much cheaper and equally effective from a learning standpoint? Or do we listen to the qualitative data and try to figure out a different method for presenting messages that users do not find as intrusive, perhaps by letting them decide when they want their random tutor activated?

Triangulation Across Scientific Disciplines

McGrath, Martin, and Kulka (1982) discussed the dilemmas faced by researchers when choosing among research methods. They identified eight distinguishable research strategies: Laboratory experiments, Experimental simulations, Field experiments, Field Studies, Computer Simulations, Formal Theory, Sample Surveys, and Judgment Tasks.

They illustrated the relations among these research methods by laying them out as pie slices in a circle. The eight methods are classified as occurring in "natural settings, in contrived or created settings, independent of natural settings or methods in which no observation of behavior is required" (p. 73). So, for example, field experiments and field studies are classed together as "settings in natural systems," whereas formal theory and computer simulations are "methods that do not require direct observation of behavior." Methods are also classified as obtrusive or unobtrusive and as universal or particular. They pointed out that, other things being equal, the researcher hopes to maximize three mutually conflicting goals:

> A. generalizability with respect to populations, B. precision in control and measurement of variables related to the behaviors of interest, and C. existential realism, for the participants, of the context within which those behaviors are observed. But alas, ceteris is never paribus, in the world of research. The very choices and operations by which one can seek to maximize any one of these will reduce the other two, and the choices that would "optimize" on any two will minimize on the third. Thus, the research strategy domain is a three-horned dilemma, and every research strategy either avoids two horns by an uneasy compromise but gets impaled, to the hilt, on the third horn; or it grabs the dilemma boldly by one horn, maximizing on it, but at the same time "sitting down" (with some pain) on the other two horns. (p. 74)

This is the crux of the problem. Field studies maximize "existential realism" but suffer with respect to generalizability and precision in control. Laboratory experiments maximize precision of measurement but suffer with respect to generalizability and existential realism. Experimental simulations and field experiments also fit between these, but they too involve compromises. In Mackay and Fayard (1997), we described a set of research studies at the C.E.N.A., the French center for studies in air traffic control. The studies included biological analyses of sleep patterns of controllers, laboratory experiments of different user interface strategies, computer simulations of new tools used by controllers, cognitive models of air traffic control's activities, and ethnographic studies of controllers at work. We argued that triangulation across these different research methods provides a deeper understanding of the problems faced by controllers and leads to better design solutions. Even so, triangulation across scientific research methods is not always obvious. We need to develop methods for reporting research results that make it easier to make cross-disciplinary comparisons.

Triangulation Across Science and Design Disciplines

HCI needs a sophisticated model of triangulation: comparing quantitative and qualitative results within individual studies, performing different kinds of experiments to test the same phenomenon, and comparing results from studies derived from different scientific disciplines. However, even this is not enough. All of these forms of triangulation assume a shared scientific context. Yet HCI has another very important component: design.

By designers, I do not mean people trained in computer science or psychology. I mean people with academic training in a design discipline, whether it is graphic design, architecture, or typography. The concept of building upon previous work exists within design disciplines, but in a manner very different from the sciences. Designers are trained how to see (or hear) as well as how to produce. Design students are taught design rules and are then evaluated on how well they break them. Design mixes craft and artistic sense; designers learn by exposing themselves to a variety of ideas, good and bad. Creating quality interactive software has an important design component: A good user interface is more than the lack of usability problems. For designers, triangulation concerns the use of multiple techniques for creating new artifacts, rather than evaluating them.

HCI is a strange field. The artifacts we study are not natural phenomena in the ordinary scientific sense; they are artifacts created by people. We may study the artifacts themselves, but more commonly, we study the interaction between people and these artifacts. We can benefit greatly from techniques borrowed from other scientific fields, but we must reexamine them and, in some cases, recreate them. Triangulation is probably the only realistic solution for dealing with different underlying assumptions and clashing paradigms. Yet we have much to learn about how best to triangulate within and across scientific and design disciplines.

Conclusion

Gray and Salzman have touched the tip of an iceberg. Their detailed analysis of usability studies has highlighted the need for a better research foundation for HCI. The HCI community has the opportunity to address and discuss the underlying research context and peer review process. If we add the concept of triangulation within and across scientific and design disciplines, we can facilitate communication among researchers and designers and improve the scientific basis for HCI.

References

Bailey, R. W., Allan, R. W., & Raiello, P. (1992). Usability testing vs. heuristic evaluation: A head-to-head comparison. *Proceedings of the Human Factors Society 36th Annual Meeting,* 409–413. Santa Monica, CA: Human Factors Society.

Cook, T., & Campbell, D. (1979). *Quasi-experimentation: Design and analysis issues for field settings.* Boston: Houghton-Mifflin.

Mackay, W. E., & Fayard, A. L. (1997). HCI, natural science and design: A framework for triangulation across disciplines. *Proceedings of the DIS'97 Conference on Designing Interactive Systems,* 223–234. New York: ACM.

McGrath, J., Martin, J., & Kulka, J. (1982). *Judgment calls in research.* Newbury Park, CA: Sage.

Molich, R., & Nielsen, J. (1990). Improving a human–computer dialogue. *Communications of the ACM, 33,* 338–348.

Runkel, P., & McGrath, J. (1972). *Research on human behavior: A systematic guide to method.* New York: Holt, Rinehart & Winston.

10. On Simulation, Measurement, and Piecewise Usability Evaluation

William M. Newman

The author is a Principal Scientist and manager of the Collaborative and Multimedia systems group at the Cambridge laboratory of Xerox Research Centre Europe; he is involved in research investigating ways of establishing evaluation parameters for interactive systems. Address: Xerox Research Centre Europe, 61 Regent Street, Cambridge CB2 1AB, UK. E-mail: william.newman@xrce.xerox.com.

Usability evaluation is, as Gray and Salzman tell us, a confusing topic. Their article painstakingly and comprehensively explains away one source of confusion, namely the conflicting recommendations published in the HCI literature about the relative merits of usability evaluation methods (UEMs). Their review of these experimental comparisons is remarkable for its thoroughness and clarity, for the plethora of flaws it identifies in the experiments, and for the range of issues it raises. Gray and Salzman appear nevertheless to have resisted almost every opportunity to explore these issues, as if determined to practice what they preach and avoid any claims that their data cannot support. They therefore present us with an overabundance of possible directions for further exploration and reflection.

In the end, I have found it interesting to reflect on two basic elements of testing and evaluation—*simulation* and *measurement*—and to consider how these contribute to the state of affairs described in the article. Simulation is fundamental to evaluation because, as Gray and Salzman point out, during design we cannot create the exact real-world conditions in which systems will perform. Instead we must simulate these conditions as best we can. Measurement is also fundamental, because it tells us whether the design is achieving its targets. I would like to have seen more reference to these elements of evaluation in the article, because I am sure the authors have important things to say about them. After all, Gray himself took part in the pioneering Project Ernestine effort to simulate a real-world system and to measure its performance, both analytically and empirically (Gray, John, & Atwood, 1993). To anyone familiar with the meticulous evaluation methods of Project Ernestine, some if not all of the UEMs discussed by Gray and Salzman must indeed look like damaged merchandise. The inherent weaknesses in the UEMs come to light when we examine how each method deals with simulation and measurement.

Simulation in Usability Evaluation

Any method for testing a design must rely to some extent on simulating the circumstances of use. Gray and Salzman hint at this central issue in evaluation when they discuss external validity. They refer several times to the unattainable "truth" about the real-world use of systems. This truth can be observed only when the real system is operated by its real users, performing real activities under real external conditions. During design, none of these aspects of the truth is readily available to evaluators. Hence there is a need to simulate the truth in order to evaluate.

A number of simulations are in fact involved in making predictions about an interactive system's use in the real world. The system itself must be simulated, if not by a working prototype, then by a mock-up such as a sequence of screen layouts. The performance of tasks by the user will need to be simulated too, based on what the evaluators know about real-world task goals and methods. The eventual user or users may not be available to take part in evaluations and so will need to be simulated by stand-ins, possibly by the evaluators themselves. Various aspects of the environment surrounding the system and its users (physical, social, etc.) may be simulated to a varying degree, or simply ignored. Each of these simulations can thus be conducted in a variety of ways, and each variation affects the simulation's completeness and accuracy, with a cumulative effect on the evaluation as a whole.

When evaluation is based on simulations that are inaccurate or incomplete in these various respects, the results can be misleading. The UEMs discussed by Gray and Salzman involve simulations that are flawed in a number of respects. For example, in most of the walkthrough UEMs discussed, simulations of technology must be based on written descriptions or sequences of screen layouts and will rely for accuracy on the interpretation skills of the evaluator. These same UEMs tend to rely on the evaluator to impersonate the user, rather than involve people who will use the system in real life. Of course many of the systems that are evaluated in this way support generic applications such as office work or voice-based interaction. Right or wrong, designers and evaluators of these systems are prone to adopt the attitude that "since I could use it, I'm competent to design it and to test it too" (Pheasant, 1991). This attitude is invalid if the application domain is a specialized one like nursing, teaching, or crime prevention. Yet the dangers inherent in letting human factors experts or software engineers conduct walkthrough-based evaluations of specialized applications are rarely spelled out.

User Testing as the Gold Standard

There is of course a fall-back approach to evaluation, user testing, which corrects many of these flaws in simulation. Throughout their article Gray and Salzman treat user testing as a kind of gold standard against which to judge other UEMs, even though user testing itself obviously falls short of simulating the truth. For example, they emphasize that user testing measures performance directly, whereas walkthrough methods involve mapping from problems to performance. They consider walkthrough methods more vulnerable to causal construct invalidity, and they question the success of these methods in finding problems. They appear to find no fault with the results of user testing, and I find this surprising and indeed worrying. User testing can deliver misleading results even when carefully carried out, and documented cases illustrate this.

One such case emerged after the tragic shooting down of Iran Air flight 655 in the Persian Gulf by the USS Vincennes in July 1988. In the aftermath, attention focused on the usability of the ship's on-board Aegis antiaircraft weapon system. The system's large-screen displays had been subjected to extensive prior user testing, when exercises were staged involving as many as 40 raids against an Aegis-equipped ship. Nevertheless, during the Persian Gulf incident the system failed to provide the commanding officer with correct classification and altitude information. Transponder information was picked up indicating that the aircraft was an F-14 fighter, and this was posted on the display. Shortly afterward, the aircraft was reported as descending through 7,800 ft and gaining speed, although in reality it was climbing at constant speed through 12,000 ft. As a result the Iranian aircraft was mistakenly classified as hostile and shot down, with the loss of 290 lives (U.S. Congress, Senate Committee on Armed Services, 1988).

Another case is mentioned in the reports on Project Ernestine, in which a telephone company investigated the performance of two telephone-operator workstations, one of which was being considered as a replacement for the other. The new workstation incorporated a number of improvements to the display, keyboard, operating procedures, and mainframe communications. It had been thoroughly user tested during its design, with the aid of appropriate industry-standard human factors evaluation techniques (Atwood, Gray, & John, 1996). As a result, the new workstation was expected to save up to 2 sec on every call handled. However, the Project Ernestine field trial, with its real users, real tasks, and real work conditions, showed otherwise: The new workstation averaged two thirds of a second slower per task performed. On this basis, the telephone company decided not to purchase the new workstation.

These two cases illustrate the need to pay attention to the quality of the simulations employed in user testing and especially the simulations of users' activities by means of *selected representative tasks*, such as establishing whether a target is a military or commercial aircraft or entering a customer's credit-card number. What kinds of inaccuracies can creep in when users' activities are simulated in this way? For a start, we should consider the constraints under which user testing is often carried out and the difficulties these create for those choosing the tasks. Increasingly, interactive systems are developed to be used in conjunction with other systems rather than in isolation, but for practical reasons they must often be user tested in isolation. Problems in performing more complex tasks may then go undetected.

For example, the Aegis system's large-screen displays were designed for use in parallel with a number of small tabular computer read-out (CRO) consoles in the same room. The CRO consoles were showing current target data, including bearing, range, speed, altitude, and IFF status (transponder classification). An incorrect IFF reading was taken for Flight 655 when the radio receiver picked up a transponder signal from an F-14 on the ground in Iran. After this was posted on the large-screen display, the radio receiver was inadvertently left at this range setting for a crucial 90 sec, and the incorrect display label remained unchanged (Adam, 1988). Meanwhile the CRO operator was reading altitude data from the CRO display and relaying it to those watching targets on the large-screen display; it has been suggested that the operator may have started reading range data as altitude (U.S. Congress, Senate Committee on Armed Services, 1988). We do not know whether the designs for these two components of Aegis (the large-screen displays and the CRO consoles) had been evaluated separately, but this would have been consistent with contemporary accounts of Aegis evaluations, all of which focus on testing individual components of the system (Broyles, 1990; Heasly, Dutra, Kirkpatrick, Seamster, & Lyons, 1993).

Users' activities may involve interacting with people as well as systems: doctors with patients, bank staff with customers, telephone operators with callers. These "passive users" are rarely involved in user tests; instead a set of benchmark tasks is again chosen that involves only the "active user" and the system. Only a subset of the user's activities is then being simulated—for example, the telephone operator's interactions with the system but not with the caller. There is evidence that user-testing errors of this exact kind were made during the design of the new workstation later tested by Gray et al. (1993) during Project Ernestine (J. C. Thomas, personal communication, September 2, 1997). It would have been possible for crucial performance problems, such as an increase in the time to enter data provided by the customer, to go undetected. Ultimately these kinds of performance problems outweighed the benefits gained by the workstation redesign (Gray et al., 1993).

Mistakes such as these will be avoided only if evaluators take a broader look at users' activities. It is not enough simply to look for frequently performed tasks; the overall structure of the work needs to be understood, after which tasks can be selected that influence the performance of significant units of work. For example, if the system is to support a family practitioner, studies should focus at the very least on gathering data on a sequence of consultations and on recurring structures within consultations, and individual tasks should then be selected only if their contribution to the whole consultation is understood. Then we might see fewer situations of the kind that pertains in the United Kingdom, where family doctors make notes on paper during the consultation only to type them into their computers between each consultation and the next (Greatbatch, Luff, Heath, & Campion, 1993). We might also see more evaluations as accurate as Project Ernestine's.

Measurement in Evaluation

Simulation provides a context within which measurements can be taken, making possible evaluation of the performance of the system as a whole. Problems in simulation tend to compound difficulties in taking measurements, however, and evidence of this can be seen in some of the UEMs discussed by Gray and Salzman.

With the focus of these UEMs being on problem detection, it is very hard for the reader of Gray and Salzman's article to discern whether evaluators are considering performance issues or just functional problems such as endless command loops. Anyone who has conducted heuristic evaluations or walkthroughs will know, however, that they detect large numbers of performance problems. One of the reasons for this, as Gray and Salzman point out, is that all such UEMs rely on scenarios that are either provided to evaluators at the outset or constructed by them on the fly. In walking through these scenarios, evaluators become aware of design problems through their impact on scenarios' performance—for example, hard-to-hit targets, or poorly chosen labels that will hinder exploratory learning. The stream of problems generated by these UEMs often contains large numbers of suggestions to improve the performance of various parts of the system.

In contrast, when measurements are taken during user testing they provide precise information about the performance of tasks, but this information is of little help to the designer. It must be mapped back into a set of contributing design factors in the manner described by Gray and Salzman. For example, if a particular task is performed conspicuously slowly, the designer needs to know that the slow appearance of a dialogue box is a major contributor. Ultimately it is this localized information that matters to the designer; the specific user task that exposed the problem may be of little interest.

If indeed most UEMs tend to generate a stream of problems relating just to specific parts of the system—and it would be interesting to know if this is true—the designer is unlikely to complain, for designers tend to devote their attention to altering parts of the system as a strategy toward overall improvement (Adelson & Soloway, 1985). Our main worry should not be whether designers will find these UEMs useful, therefore, but whether the UEMs will be effective in improving the overall performance of the system. This will mean gaining an understanding of what is meant by "overall performance." I cannot tell from Gray and Salzman's article whether they appreciate this point; however, they do talk about the need to evaluate in terms of *outcomes of interest* toward the very end of their article.

When there are large performance gains to be obtained, standard methods of empirical performance measurement based on multiple tasks can be very effective. Thus the sets of benchmark tasks used by Sullivan (1996) in evaluating the Windows 95 user interface and by Burkhart, Hemphill, and Jones (1994) in evaluating their telecommunications module, enabled them to estimate overall performance improvements of roughly 50% and 35%, respectively. However, when performance improvements are relatively small, sets of benchmark tasks can give unreliable measures. It becomes necessary to know how to weight the tasks in terms of their relative frequency and of course to take into account (for reasons given previously) other tasks not involving just the user and the system. If this is not done, we risk a repeat of the new workstation's design in the Project Ernestine story—that is, basing performance measures on a partial set of tasks, and ignoring other tasks whose performance is adversely affected by the system's design.

Conclusion: Beyond Piecewise Evaluation

I have enumerated a number of problems with today's evaluation methods; all of them are to some extent instances of taking *piecewise* approaches to evaluation. The primary source of this fragmentation lies in simulation of the user's activities by means of sets of separate tasks or scenarios. During evaluation each of these is performed or walked through more or less independently. Direct measurements can at most be applied piecewise to the entirety of each of the tasks. However, to provide designers with useful feedback, evaluators tend to focus even more narrowly on individual performance problems and the system components to which these problems relate. The result is piecewise delivery of evaluation results to the designer. A consequent risk is that the changes made by designers to correct the individual problems will not add up to improving the system's performance during real-world use.

The only way I can see to overcome these problems is to take more of an *overall* view of the system and its use: to understand overall usage structures and patterns, to measure units of activity that are critical to overall performance, and to assist designers in making changes that have a positive overall impact on the system's performance. To help with the last of these, Gray and Salzman advocate research aimed at understanding how intrinsic properties of a system affect payoff performance, and this is certainly needed. Studies that elicit the overall structure of the work, conducted via a combination of ethnographic and task-analytic methods, would also be helpful. These might enable evaluators to build more reliable simulations of users' activities, with less risk of overlooking important passive users and secondary systems. Such studies would help in identifying the *critical parameters* of interactive applications—that is, parameters that can become established as a basis for measuring how well the system serves its purpose and for comparing one design with another (Newman, 1997).

Documented design cases suggest that the availability of such critical parameters may help designers to create systems that achieve overall performance improvements. Designers are able to maintain a focus on the parameters in question and to limit their changes to those that do in fact make a positive difference. For example, Alm, Todman, Elder, and Newell (1993) set a goal of designing a system to enable speech-impaired people to increase their rate of conversation in words per minute and were able to take the design through to a conclusion that demonstrated a significant improvement. On a smaller scale, the study by Maclean, Young, Bellotti, and Moran (1991) of a team of designers designing a "fast automated teller" illustrated the concern for performance that comes with this kind of design goal.

If critical parameters were more widely known to designers, this style of design might become more common, leading to more frequent success in improving performance overall. As a further consequence, we would perhaps need to deal less often with damaged merchandise in the form of poorly performing systems and unreliable methods for usability evaluation.

Acknowledgments

I am most grateful to Alice Wilder-Hall and Lisa Alfke, of the Xerox PARC Information Center, for their considerable help in collecting materials for this commentary. I also thank Alex Taylor for his suggestions for changes to the final draft.

References

Adam, J. A. (1988, November). Fixes to Aegis system recommended by Navy (Suppl.). *IEEE Spectrum, 25,* 1–2.
Adelson, B., & Soloway, E. (1985). The role of domain experience in software design. *IEEE Transactions on Software Engineering, SE-11,* 1351–1360.

Alm, N., Todman, J., Elder, L., & Newell, A. F. (1993). Computer aided conversation for severely impaired non-speaking people. *Proceedings of the InterCHI'93 Conference on Human Factors in Computing Systems*, 236–241. New York: ACM.

Atwood, M. E., Gray, W. D., & John, B. E. (1996). Project Ernestine: Analytic and empirical methods applied to a real-world CHI problem. In M. Rudisill, C. L. Lewis, P. G. Polson, & T. D. McKay (Eds.), *Human–computer interface design: Success stories, emerging methods and real-world context* (pp. 101–121). San Francisco: Morgan Kaufmann.

Broyles, J. W. (1990). AEGIS status-display formats: Tradeoff studies. *Proceedings of the Department of Defense Human Factors Engineering Technical Group*, L–7. Washington, DC: U.S. Department of Defense.

Burkhart, B., Hemphill, D., & Jones, S. (1994). The value of a baseline in determining design success. *Proceedings of CHI'94 Human Factors in Computing Systems*, 386–391. New York: ACM/SIGCHI.

Gray, W. D., John, B. E., & Atwood, M. E. (1993). Project Ernestine: Validating a GOMS analysis for predicting and explaining real-world task performance. *Human–Computer Interaction, 8*, 237–309.

Greatbatch, D., Luff, P., Heath, C., & Campion, P. (1993). Interpersonal communication and human–computer interaction: An examination of the use of computers in medical consultations. *International Journal of Interacting with Computers, 5*, 193–216.

Heasly, C. C., Dutra, L. A., Kirkpatrick, M., Seamster, T. L., & Lyons, R. A. (1993). A user-centered approach to the design of a naval tactical workstation interface. *Proceedings of Human Factors and Ergonomics Society 37th Annual Meeting*, 1030. Santa Monica, CA: Human Factors and Ergonomics Society.

MacLean, A., Young, R. M., Bellotti, V. M. E., & Moran, T. P. (1991). Questions, options and criteria: Elements of design space analysis. *Human–Computer Interaction, 6*, 201–250.

Newman, W. M. (1997). Better or just different? On the benefits of designing interactive systems in terms of critical parameters. *Proceedings of the DIS'97 Conference on Designing Interactive Systems*, 239–245. New York: ACM.

Pheasant, S. (1991). *Ergonomics, work and health*. Basingstoke, England: Macmillan.

Sullivan, K. (1996). The Windows 95 user interface: A case study in usability. *Proceedings of CHI'96 Conference on Human Factors in Computing Systems*, 473–480. New York: ACM.

U.S. Congress, Senate Committee on Armed Services. (1988). *Investigation into the downing of an Iranian airliner by the USS Vincennes* (No. 90–353). Washington, DC: U.S. Government Printing Office.

HUMAN-COMPUTER INTERACTION, 1998, Volume 13, pp. 325–335

Repairing Damaged Merchandise: A Rejoinder

Wayne D. Gray and **Marilyn C. Salzman**
George Mason University

Our goal in writing "Damaged Merchandise?" (DM) was not to have the last word on the subject but to raise an awareness within the human–computer interaction (HCI) community of issues that we felt had been too long ignored or neglected. On reading the 10 commentaries from distinguished members of the HCI community, we were pleased to see that they had joined the debate and broadened the discussion. Subsequently, we were somewhat torn by how to proceed. Our first thought was to respond point by point, commentary by commentary. However, we refrain from addressing many specific issues here, as a full discussion would involve an article at least as long as DM. Instead we focus on a few important themes that emerged throughout our article and the ensuing discussion:

- What is usability, how do we measure it, and what do we need to know about our usability evaluation methods (UEMs)?
- Why do we find ourselves where we are?
- What is the role of experiments versus other empirical studies in HCI? Are there common issues in the design of empirical studies?
- How do we judge the value of a study?
- Where do we go from here?

1. WHAT IS USABILITY, HOW DO WE MEASURE IT, AND WHAT DO WE NEED TO KNOW ABOUT UEMs?

After completing our review of validity problems with the UEM studies (Section 5 of DM), we began our Observations and Recommendations (Section 6) by stating our belief that "the most important issue facing usability researchers and practitioners alike [is] the construct of usability itself" (p. 238). We argued that we need to broaden our definition of usability so that it goes beyond the problem-counting approach epitomized by the papers reviewed in DM. Among our commentators, Mackay, McClelland, Monk, Oviatt, and Newman took up and elaborated this issue.

Mackay urges the use of multiple measures to triangulate on what is meant by usability. McClelland points out that "usability is not one thing" and that it is an evolving construct. In addition to traditional usability issues (objective performance, subjective impressions, safety, and learning), we may also need to consider factors such as pleasure of use when evaluating the strengths and weaknesses of our UEMs. Oviatt presents a similar argument, encouraging the HCI community to consider more than our traditional measures of usability when examining UEMs and to rely on triangulation to pinpoint each UEM's strengths and weaknesses.

Newman's commentary touches on a number of key points that deserve careful consideration. First, he faults us for a sin of omission—namely, finding "no fault with the results of user testing"—and ties this into an excellent discussion of the type of information a designer needs to have to improve a design. Although we chose not to focus on the strengths and weaknesses of user testing in DM, it was clear to us that how user testing was implemented and how its outcomes were interpreted varied greatly among the five studies reviewed. In fact, in Section 6.1 we warned that it is difficult to determine what interface features cause which usability test problems, that the "problems-to-features" mapping "cannot be assumed and the links must be carefully forged." As Monk points out, this is an issue for which experimentation can be particularly useful. Another useful mechanism for accomplishing this goal may be simulation, an argument made by both Carroll and Newman.

Second, we believe Newman is correct in asserting that designers need more than measures of "overall performance" as they are seldom truly interested in the tasks performed in the usability lab. Designers are not interested in "the performance of tasks" but in how the interface affects that performance. "If a particular task is performed conspicuously slowly, the designer needs to know that the slow appearance of a dialogue box is a major contributor" (Newman, this issue). This sentiment echoes John and Mashyna's (1997) concerns with attempts to classify usability problems into a small number of categories. They maintain that developers need to know the specific problem (e.g., a problem with an item in a particular menu) and not the general one (e.g., "there are menu problems" or "speak the users' language").

Third, we endorse Newman's argument that we need to go beyond "piecewise approaches to evaluation" to identify "the critical parameters of interactive applications." We believe that one way to do so is to combine user testing with careful, fine-grained task analyses such as those that can be accomplished via GOMS (John & Kieras, 1996a, 1996b). An excellent example of this is Franzke's study of the use characteristics of different software packages (Franzke, 1994, 1995). Although such studies are currently painstaking to perform, a large part of the pain is caused by the absence of off-the-shelf tools to automate the process of putting user data into correspondence with the various components of a fine-grained procedural task analysis. Perhaps this is an area in which the tool-building component of the HCI community can come to the aid of those interested in measurement and methodology.

2. WHY DO WE FIND OURSELVES WHERE WE ARE?

In exploring the question of why some of the best industry laboratories in our field could have produced research with the flaws delineated in DM, our commentators seem to be of two minds. The first camp argues, as do Jeffries and Miller, that valid research is not interesting research. They pose a dichotomy between doing valid experiments and experiments that have *ecological validity*. The second camp (which is well represented by Lund and Mackay) argues that research is valued within corporations based on its *face validity* rather than its methodological purity. Thus, as Lund states, "it is almost axiomatic, therefore, that much corporate research undertaken in this environment will have limited validity." When practitioners review conference and journal submissions, they are heavily influenced by the values of the corporate culture within which they work and look for "content that *appears* to provide value" [italics added]. Mackay sees this problem as exacerbated in HCI because it is a field that draws on people with diverse backgrounds, in which experts in one area may lack training in other areas such as experimental design. Whereas diversity is a strength of the field, aligning reviewer expertise with the conceptual, methodological, and statistical issues raised by various research papers is a challenging endeavor.

We believe the distinction between face validity and ecological validity is important. The former *appears* good, the latter *is* good; the former *is* easy, the latter *is* hard; and, unfortunately, *the former is often mistaken for the latter*. When empirical techniques, experimental or otherwise, are applied to address difficult real-world problems, it is much easier to do something that *appears* good, rather than something that *is* good.

This tendency toward face validity may be exacerbated (as suggested by Lund) by the absence of a focus on methodological issues within the HCI-oriented, academic research community. He points out that "the absence of comparative academic research in this area is noticeable and distressing" and surmises that gatekeeping with respect to the conference review process may have failed in part due to the "relatively limited academic interest in the topics of methodologies."

3. WHAT IS THE ROLE OF EXPERIMENTS VERSUS OTHER EMPIRICAL STUDIES IN HCI? ARE THERE COMMON ISSUES IN THE DESIGN OF EMPIRICAL STUDIES?

DM focused on experiments and problems with experimental design for the simple reason that the five studies we reviewed were cast by their authors as experimental studies. It was problems with the use of the experimental method in these studies and their acceptance by the HCI community that motivated DM. DM should not be interpreted either as advocating experiments to the exclusion of other empirical techniques or as maintaining that other techniques do not need to be concerned with the types of validity discussed in DM.

Although experiments were our focus, throughout DM we included small statements of our regard for other empirical methods. For example, our conclusion section begins, "The multitude of empirical methodologies is a strength of the HCI research community" and in Section 3 we acknowledge the "plethora of alternatives" to experiments. Apparently, these small statements were not enough. In her commentary, John expresses her fear that our "unremitting focus on problems with experimental design" might make it easy for some readers to conclude that we believe that all HCI research must be cast as experiments. Unfortunately, this "unremitting focus" has apparently mislead Carroll, who states that we advocate experiments as "the royal road to causal analysis." We do not.

As Cook and Campbell (1979) discussed in Chapter 2 of their book, experiments enable the researcher to draw strong inferences concerning causality. (Indeed, we suspect that this is why so many researchers attempt to cast their work in experimental form despite the plethora of alternatives.) This power comes at a price—experiments do not degrade gracefully. Although there may be a continuum of experimental designs ranging from well designed to poorly designed, there is not a concomitant continuum of inferences, from strong to weak, that can be drawn from such designs. At some point along the design continuum, valid inferences no longer can be made. We need not only to value the strength of the inferences that an experiment permits, but to understand that small problems in experimental design can have large effects on what we can legitimately conclude. Flawed experiments simply do not enable us to draw valid inferences.

When circumstances do not permit a valid experiment to be designed, we urge researchers to consider alternatives to experiments. One alternative is to do as Mackay recommends–to go beyond Chapter 2 to the core of Cook and

Campbell's (1979) excellent book. Chapters 3 and beyond are devoted to *quasi-experimental design*: the detailed consideration of how to design reliable and valid empirical studies when valid experimental designs are not an option. We urge the reader not to be bothered, as Carroll apparently is, by Cook and Campbell's "quirky language." These researchers were not writing for experimental psychologists or for computer scientists. They were addressing the public policy-oriented program evaluation community (i.e., large social and education programs, not computer programs). It would be a mistake for the discipline of HCI to ignore the lessons embodied in that book simply because its language comes from another discipline.

Other alternatives to experimental design mentioned by our commentators include ethnographic studies, simulation studies, case studies, diary studies, critical incident techniques, task analytic methods, and verbal protocol analyses. Such alternatives can be used to study issues and relations that cannot easily be brought under experimental control. For example, if our goal is to understand how UEMs are applied in practice, some of the alternatives (e.g., ethnographic studies, case studies, etc.) cited previously might be more effective than experimentation. If our goal is to better define the usability construct, simulation studies and task analytic methods might be particularly useful. Finally, if our goal is to understand how successful UEMs are at discovering rare but critical events, we might turn to the critical incident technique, which is a form of task analysis (Kirwan & Ainsworth, 1992).

In turning to nonexperimental forms of empirical inquiry it is necessary to remember that these are not simply sloppy experiments, but they have their own requirements of methodological rigor. For example, John rightly criticizes our statement that

> If Jeffries et al. (1991) had been cast as a case study (and appropriate changes made throughout), the paper would have provided a snapshot of the trade-offs facing Hewlett Packard in deciding how to do usability analyses in the late 1980s.

John is correct. Recasting Jeffries, Miller, Wharton, and Uyeda (1991) as a case study would not alleviate the validity issues we raised. As John's comment suggests, even techniques that do not lend themselves to rigorous statistical analysis must face concerns with internal validity, construct validity, and external validity.[1]

---

1. Excellent discussions of methodological rigor in nonexperimental empirical studies include Yin's (1994) discussion of case studies and Ericsson and Simon's (1993) or van Someren, Barnard, and Sandberg's (1994) discussion of methodological issues for verbal protocol analysis.

In choosing an empirical technique, experimental or otherwise, several factors need to be carefully considered. First, different empirical techniques support stronger or weaker inferences regarding causality. Second, there is no best empirical technique, only a best technique given the question and circumstances. Third, there may well be a trade-off between power and the breadth of questions that are asked so that techniques such as case study or verbal protocol analysis may allow weaker conclusions to be made about a wider range of issues than a comparable experimental study. Fourth, our statements about the strength of causal inferences that various techniques permit apply only to single studies. A well-designed series of, for example, case studies, may well support causal inferences as strong as any permitted by experimentation. Fifth, along with all of our commentators who took a stance, we believe in empirical pluralism or triangulation. Difficult questions, especially those confronting HCI, are unlikely to be adequately addressed using any one technique. Sixth, triangulation involves more than simply using different empirical techniques to study the same issue. It requires that the results from these divergent empirical approaches be combined and integrated into powerful analytic frameworks. Such frameworks will provide a source for practitioners to draw on and will define the gaps and issues that researchers should explore. Undoubtedly, the resulting frameworks will be as diverse as the needs of the practitioner and researcher communities. However, if done well, the frameworks will be partially overlapping and mutually referencing. For issues concerning individual cognition, emerging frameworks can be found in the form of cognitive architectures and their use by the HCI research community in computational cognitive modeling (see, e.g., Gray, Young, & Kirschenbaum, 1997).

4. HOW TO JUDGE THE VALUE OF A STUDY?

Many of our commentators weigh in on the issue of how to judge the value of a study. Several take a narrow perspective, focusing on the problems involved in gaining access to large numbers of usability specialists or software engineers. A pair of commentators complain that DM misjudged the value of the study they had conducted. Finally, some of our commentators chose to comment on the place of individual studies in the context of the larger research enterprise. In this section, we briefly comment on the first two approaches and elaborate on the third.

A narrow focus on statistical conclusion validity led several of our commentators to discuss the difficulty of obtaining large samples of the type of participants needed for UEM research. Although gaining access to the right type of participants is important, valid and useful results can be achieved with a limited participant pool. In fact, the most serious problems in inferring cause-and-effect in the studies we reviewed did not stem from lack of access to or low numbers of the right people; rather, the problems stemmed from other features of the experimental design. Three of the studies we reviewed had adequate sample sizes but had important problems with other validity issues. Likewise, for the two studies with lowest sample sizes, it is fair to say that their overall validity would not have been improved by a tenfold increase in sample size. Although in experimental design, as in much of life, more is generally better, having a large sample size does not guarantee success, and, as witnessed by several studies mentioned in DM, a low sample size does not guarantee failure.

Jeffries and Miller maintain that we misjudged the value of Jeffries et al. (1991) by failing to appreciate that the value of a study depends on understanding "*what* kind of study is being done, *why* it's being done, and *whether* it has a chance of answering the questions it poses" [italics added]. Although this strikes us as a reasonable set of principles, apparently we disagree with Jeffries and Miller as to how these principles apply to their study. Specifically, for Jeffries et al., we evaluated what and why based on the questions posed in their introduction and the conclusions drawn in their Discussion section. The whether was judged based on their Methodology and Results section. We stand by our statements that the questions posed cannot be addressed by the study as conducted and that the conclusions drawn are not warranted by the results obtained. We refer readers to DM's Appendix, in which we carefully delineate why the conclusions drawn by Jeffries et al. are not supported by their study.

Other commentators, including Carroll and Karat, take a broad perspective and speak about the difficult challenge of balancing rigor and relevance in HCI research. We share this perspective. Finding this balance requires the careful identification of research questions and empirical methods to answer those questions. Such a process involves satisficing not optimizing. Our choices must be tempered by the resources we have available. First, we must limit the scope of our questions so that they are important but answerable. Second, we must design a study or series of studies that can adequately address those questions. When resources such as participants are limited, we need to look for ways to utilize those resources effectively. This may mean using repeated measures in an experimental design framework or using nonexperimental techniques such as case studies. Third, we need to recognize that each method has its own rules of inference and its own standards of methodological rigor. When we chose to cast our study in an experimental framework, we need to carefully construct our design so that it has statistical conclusion, internal, construct, and external validity. These issues cannot be ignored when we turn to nonexperimental techniques. In fact, we should use our knowledge about validity to guide our selection of appropriate methods. When a method's validity is violated, inferences cannot be drawn.

5. WHERE DO WE GO FROM HERE?

Among the majority of our commentators, the consensus on what the field of HCI should do to improve our knowledge of usability and UEMs can be summarized by saying that "we just have to think bigger" (B. E. John, personal communication, February 13, 1998). In DM, we argued that no single measure of usability could establish the reliability and validity of a UEM and that multiple converging measures (Section 6.1) were needed. We join with several of our commentators (Mackay, Oviatt, John, Newman, Monk, Carroll, and Karat) in extending this argument to include multiple converging techniques. Triangulation is critical to the advancement of our understanding, whether our focus is on UEMs or on other issues of HCI. In this issue, Mackay provides a thought-provoking discussion on how triangulation can help improve the conference and journal review process as well as facilitate more effective studies of UEMs. John describes one use of triangulation to discover the strengths and weaknesses of UEMs.

An important component of thinking bigger is having a research agenda that extends beyond one's current study. There are many ways to characterize UEMs and many criteria by which to judge their effectiveness. Finding usability problems is central, critical, and complex in it own right. However, other factors such as when a UEM can be used, who should administer it, what types of products it is suited for, and the time and resources it demands are also very important (John, this issue; Olson & Moran, 1996). As Karat points out, UEMs have different purposes, making it difficult, and perhaps undesirable to compare them solely by the type of usability problems they find.

Many of the points made in DM, when taken together with many of the points made by Oviatt, Newman, John, McClelland, Mackay, Monk, and Karat, may be seen as laying out a research agenda for repairing damaged merchandise. To be useful to practitioners, research needs to

- Develop definitions and measures of usability [construct validity issues] that are applicable in a variety of context (e.g., real-time or safety-critical systems, consumer appliances, point-of-sales issues, traditional office automation, etc.).
- Address the interaction between UEM and the skills required to apply the UEM [an external validity of persons issue].
- Determine which UEMs work best at what stage of the design [a construct validity issue concerning both mono-operation and mono-method biases].

- Determine which UEMs work best with what types of software [another construct validity issue, also concerning both mono-operation and mono-method biases].
- Determine in which settings (e.g., lab vs. end-user's setting) various UEMs can be best applied [an external validity of setting issue].
- Develop UEMs that avoid piecewise evaluation by considering the usability of software in its context of use [a causal construct validity issue as well as an external validity of setting issue].

To achieve this agenda, it is clear that multiple measures of usability are essential, that measures other than usability are necessary to understand the potential of UEMs, that experimentation must be supplemented by alternative methods, and that neither industry nor Academe can answer these questions alone. Of our commentators, Lund is very articulate on this last point. Speaking from a perspective informed by years of corporate experience, he points out that "the environment in which practitioners and many researchers work within corporations is unlikely to change" and suggests that mechanisms need to be created to enable the academic community to work with industry to "produce more generalizable results." "Teaming between industrial and academic researchers" is strongly endorsed by Oviatt. It should not be surprising that we concur. It is still true that as academics we have much freedom to set our own research agenda. However, this freedom is necessarily constrained by resource availability. It would be interesting and important to develop mechanisms by which university researchers could work with industry developers without the research being subverted by the corporate pressures that Lund so convincingly details.

6. CONCLUSIONS

Although this rejoinder addressed a few of the specific points raised by our commentators, for the most part we avoided considering each of the many points they made to focus on a few of the themes that emerged overall. In doing so we see this article as a contribution to an ongoing discussion of the methodology, empirical techniques, and rules of inference that should guide HCI research, rather than as a defense of DM. We look forward to being involved in these discussions with those who wish to move the field forward.

In this vein, we eschew having the last word in this first round of discussion and give that privilege to one of our commentators:

> Gray and Salzman have touched the tip of an iceberg. Their detailed analysis of usability studies has highlighted the need for a better research foundation for HCI. The HCI community has the opportunity to address and discuss the

underlying research context and peer review process. If we add the concept of triangulation within and across scientific and design disciplines, we can facilitate communication among researchers and designers and improve the scientific basis for HCI. (Mackay, this issue, p. 315)

NOTES

Acknowledgments. We give our thanks to Deborah A. Boehm-Davis for her comments on earlier versions of this article. We likewise thank our commentators for engaging us in debate and for striving to keep the tone of the discussion focused on the highest of professional purposes. We also thank our editors, Gary Olson and Thomas Moran, for highlighting the importance of the issues we raised by turning an individual submission into a special issue, for soliciting and refereeing the commentaries, and for proactively soothing the many ruffled feathers that this process inevitably entailed.

Support. The work on this article was supported by Grant IRI–9618833 from the National Science Foundation to Wayne D. Gray.

Authors' Present Addresses. Wayne D. Gray, Human Factors and Applied Cognitive Program, George Mason University, msn 3f5, Fairfax, VA 22030, USA. E-mail: gray@gmu.edu. Marilyn C. Salzman, U S WEST Advanced Technologies, 4001 Discovery Drive, Suite 370, Boulder, CO 80303, USA. E-mail: mcsalzm@advtech.uswest.com.

HCI Editorial Record. First manuscript received March 23, 1998. Revision received April 13, 1998. Accepted by Gary Olson and Thomas Moran. — *Editor.*

REFERENCES

Cook, T. D., & Campbell, D. T. (1979). *Quasi-experimentation: Design and analysis issues for field settings.* Chicago: Rand McNally.

Ericsson, K. A., & Simon, H. A. (1993). *Protocol analysis: Verbal reports as data* (Rev. ed.). Cambridge, MA: MIT Press.

Franzke, M. (1994). *Exploration and experienced performance with display-based systems* (PhD Dissertation, ICS Tech. Rep. 94–07). Boulder: University of Colorado.

Franzke, M. (1995). Turning research into practice: Characteristics of display-based interaction. *Proceedings of the CHI'95 Conference on Human Factors in Computing Systems,* 421–428. New York: ACM.

Gray, W. D., Young, R. M., & Kirschenbaum, S. S. (1997). Introduction to this special issue on cognitive architectures and human–computer interaction. *Human–Computer Interaction, 12,* 301–309.

Jeffries, R., Miller, J. R., Wharton, C., & Uyeda, K. M. (1991). User interface evaluation in the real world: A comparison of four techniques. *Proceedings of the CHI'91 Conference on Human Factors in Computing Systems,* 119–124. New York: ACM.

John, B. E., & Kieras, D. E. (1996a). The GOMS family of user interface analysis techniques: Comparison and contrast. *ACM Transactions on Computer–Human Interaction, 3,* 320–351.

John, B. E., & Kieras, D. E. (1996b). Using GOMS for user interface design and evaluation: Which technique? *ACM Transactions on Computer–Human Interaction, 3,* 287–319.

John, B. E., & Mashyna, M. M. (1997). Evaluating a multimedia authoring tool with cognitive walkthrough and think-aloud user studies. *Journal of the American Society of Information Science, 48*(9).

Kirwan, B., & Ainsworth, L. K. (Eds.). (1992). *A guide to task analysis.* Washington, DC: Taylor & Francis.

Olson, J. S., & Moran, T. P. (1996). Mapping the method muddle: Guidance in using methods for user interface design. In M. Rudisill, C. Lewis, P. G. Polson, & T. D. McKay (Eds.), *Human–computer interface designs: Success stories, emerging methods, and real world context.* San Francisco: Morgan Kaufmann.

van Someren, M. W., Barnard, Y. F., & Sandberg, J. A. C. (1994). *The think aloud method: A practical guide to modeling cognitive processes.* New York: Academic.

Yin, R. K. (1994). *Case study research: Design and methods* (2nd ed.). Thousand Oaks, CA: Sage.

Subscription Order Form

Please ❑ enter ❑ renew my subscription to

HUMAN-COMPUTER
INTERACTION
Volume 13, 1998, Quarterly

Subscription prices per volume:

Individual: ❑ $45.00 (US/Canada) ❑ $75.00 (All Other Countries)

Institution: ❑ $265.00 (US/Canada) ❑ $295.00 (All Other Countries)

Subscriptions are entered on a calendar-year basis only and must be prepaid in US currency -- check, money order, or credit card. **Offer expires 12/31/98. NOTE: Institutions must pay institutional rates.** Individual subscription orders paid by institutional checks will be returned.

❑ **Payment Enclosed**

Total Amount Enclosed $_____

❑ **Charge My Credit Card**

❑ VISA ❑ MasterCard ❑ AMEX ❑ Discover

Exp. Date_____

Card Number _____

Signature _____
(Credit card orders cannot be processed without your signature.)

Please print clearly to ensure proper delivery.

Name _____

Address _____

_____

City _____ State _____ Zip+4 _____
Prices are subject to change without notice.

Lawrence Erlbaum Associates, Inc.
Journal Subscription Department
10 Industrial Avenue, Mahwah, NJ 07430
(201) 236-9500   FAX (201) 236-0072

Ergonomics and Safety of Intelligent Driver Interfaces

Edited by
Y. Ian Noy

ERGONOMICS AND SAFETY OF INTELLIGENT DRIVER INTERFACES

edited by
Y. Ian Noy
Transport Canada
A VOLUME IN THE HUMAN FACTORS IN TRANSPORTATION SERIES

Even to the casual observer of the automotive industry, it is clear that driving in the 21st century will be radically different from driving as we know it today. Significant advances in diverse technologies such as digital maps, communication links, processors, image processing, chipcards, traffic management, and vehicle positioning and tracking, are enabling extensive development of intelligent transport systems (ITS). Proponents of ITS view these technologies as freeing designers to re-define the role and function of transport in society and to address the urgent problems of congestion, pollution, and safety. Critics, on the other hand, worry that ITS may prove too complex, too demanding, and too distracting for users, leading to loss of skill, increased incidence of human error, and greater risk of accidents.

The role of human factors is widely acknowledged to be critical to the successful implementation of such technologies. However, too little research is directed toward advancing the science of human-ITS interaction, and too little is published which is useful to system designers. This book is an attempt to fill this critical gap. It focuses on the intelligent driver interface (IDI) because the ergonomics of IDI design will influence safety and usability perhaps more than the technologies which underlie it.

The chapters cover a broad range of topics, from cognitive considerations in the design of navigation and route guidance, to issues associated with collision warning systems, to monitoring driver fatigue. The chapters also differ in intent — some provide design recommendations while others describe research findings or new approaches for IDI research and development. Based in part on papers presented at a symposium on the ergonomics of in-vehicle human systems held under the auspices of the 12th Congress of the International Ergonomics Association, the book provides an international perspective on related topics through inclusion of important contributions from Europe, North America, and Japan.

Many of the chapters discuss issues associated with navigation and route guidance because such systems are the most salient and arguably the most complex examples of IDI. However, the findings and research methodologies are relevant to other systems as well making this book of interest to a wide audience of researchers, design engineers, transportation authorities, and academicians involved with the development or implementation of ITS.

0-8058-1955-X [cloth] / 1997 / 448pp. / $89.95
0-8058-1956-8 [paper] / 1997 / 448pp. / $45.00

Lawrence Erlbaum Associates, Inc.
10 Industrial Avenue, Mahwah, NJ 07430
201/236-9500   FAX 201/236-0072

Prices subject to
change without notice.

Call toll-free to order: 1-800-9-BOOKS-9...9am to 5pm EST only.
e-mail to: orders@erlbaum.com
visit LEA's web site at http://www.erlbaum.com

USER- CENTERED REQUIREMENTS
The Scenario- Based Engineering Process
Karen L. McGraw
Cognitive Technologies
Karan Harbison
University of Texas at Arlington

Developing today's complex systems requires more than just good software engineering solutions. Many are faced with complex systems projects, incomplete or inaccurate requirements, canceled projects, or cost overruns, and have their systems' users in revolt and demanding more. Others want to build user-centric systems, but fear managing the process. This book describes an approach that brings the engineering process together with human performance engineering and business process reengineering. The result is a manageable user-centered process for gathering, analyzing, and evaluating requirements that can vastly improve the success rate in the development of medium-to-large size systems and applications.

Unlike some texts that are primarily conceptual, this volume provides guidelines, "how-to" information, and examples, enabling the reader to quickly apply the process and techniques to accomplish the following goals:
- define high quality requirements,
- enhance productive client involvement,
- help clients maintain competitiveness,
- ensure client buy-in and support throughout the process,
- reduce missing functionality and corrections, and
- improve user satisfaction with systems.

This volume clearly details the role of user-centered requirements and knowledge acquisition within Scenario-Based Engineering Process (SEP) and identifies SEP products and artifacts. It assists project personnel in planning and managing effective requirements activities, including managing risks, avoiding common problems with requirements elicitation, organizing project participants and tools, and managing the logistics. Guidelines are provided for selecting the right individual and group techniques to elicit scenarios and requirements from users, subject matter experts, or other shareholders, and ensuring engineers or analysts have the necessary skills.

Contents: Part I: *Introduction to the Scenario-based Engineering Process.* Engineering Activities and Artifacts. **Part II:** *Process & Techniques.* Planning and Managing Effective Requirements Activities. Selecting the Right Techniques. Scenario Elicitation, Analysis, and Generation. Conducting and Analyzing Interactive Observation Sessions. Conducting and Using the Interview Effectively. Defining Work Processes and Conducting Task Analysis. Eliciting and Analyzing Domain Concepts. Using Process Tracing to Analyze the Problem-Solving Process. Conducting and Analyzing Group Sessions. Evaluating and Refining Requirements.
0-8058-2064-7 [cloth] / 1997 / 392pp. / $79.95
0-8058-2065-5 [paper] / 1997 / 392pp. / $45.00

Lawrence Erlbaum Associates, Inc.
10 Industrial Avenue, Mahwah, NJ 07430
201/236-9500 FAX 201/236-0072

Prices subject to
change without notice.

Call toll-free to order: 1-800-9-BOOKS-9...9am to 5pm EST only.
e-mail to: orders@erlbaum.com
visit LEA's web site at http://www.erlbaum.com

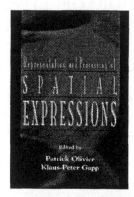

REPRESENTATION
AND PROCESSING
OF SPATIAL EXPRESSIONS

edited by
Patrick Olivier
Department of Computer Science, Aberystwyth
Klaus-Peter Gapp
Universitaet des Saarlandes, Germany

Coping with spatial expressions in a plausible manner is a crucial problem in a number of research fields, specifically cognitive science, artificial intelligence, psychology, and linguistics. This volume contains a set of theoretical analyses as well as accounts of applications which deal with the problems of representing and processing spatial expressions. These include dialogue understanding using mental images; interfaces to CAD and multi-media systems, such as natural language querying of photographic databases; speech-driven design and assembly; machine translation systems; spatial queries for Geographic Information Systems; and systems which generate spatial descriptions on the basis of maps, cognitive maps, or other spatial representations, such as intelligent vehicle navigation systems.

Though there have been many different approaches to the representation and processing of spatial expressions, most existing computational characterizations have so far been restricted to particularly narrow problem domains, usually specific spatial contexts determined by overall system goals. To date, artificial intelligence research in this field has rarely taken advantage of language and spatial cognition studies carried out by the cognitive science community. One of the fundamental aims of this book is to bring together research from both disciplines in the belief that artificial intelligence has much to gain from an appreciation of cognitive theories.

0-8058-2285-2 [cloth] / 1998 / 296pp. / $79.95
Special Prepaid Offer! $49.95
No further discounts apply.

Lawrence Erlbaum Associates, Inc.
10 Industrial Avenue, Mahwah, NJ 07430

Prices subject to
change without notice.

201/236-9500 FAX 201/236-0072

Call toll-free to order: 1-800-9-BOOKS-9...9am to 5pm EST only.
e-mail to: orders@erlbaum.com
visit LEA's web site at http://www.erlbaum.com